PENGUIN BOOKS

ENCOUNTERS WITH ANIMALS

Gerald Durrell was born in Jamshedpur, India, in 1925. In 1928 his family returned to England and in 1933 they went to live on the Continent. Eventually they settled on the island of Corfu, where they lived until 1939. During this time he made a special study of zoology, and kept a large number of the local wild animals as pets. In 1945 he joined the staff of Whipsnade Park as a student keeper. In 1947 he financed, organized and led his first animal-collecting expedition to the Cameroons. This was followed by a second expedition in 1948 and a third in 1949, this time to Guyana. He has also made expeditions to Paraguay and Argentina. He and his wife have recently been to New Zealand, Australia, and Malaya to film a TV series, *Two in the Bush*, in conjunction with a B.B.C. Natural History Film Unit. In 1958 he founded the Jersey Zoological Park, of which he is the director, and in 1964 he founded the Jersey Wildlife Preservation Trust. Gerald Durrell's other books include *The Bafut Beagles*, *My Family and Other Animals*, *Three Singles to Adventure*, *A Zoo in My Luggage*, *The Drunken Forest*, *The Whispering Land* and *Menagerie Manor* (all available in Penguins), and *The New Noah*, which is a Puffin Book.

GERALD DURRELL

ENCOUNTERS WITH ANIMALS

WITH ILLUSTRATIONS BY
RALPH THOMPSON

PENGUIN BOOKS

Penguin Books Ltd, Harmondsworth, Middlesex, England
Penguin Books Australia Ltd, Ringwood, Victoria, Australia
Penguin Books Canada Ltd, 41 Steelcase Road West,
Markham, Ontario, Canada
Penguin Books (N.Z.) Ltd, 182–190 Wairau Road,
Auckland 10, New Zealand

—

First published by Rupert Hart-Davis 1958
Published in Penguin Books 1963
Reprinted 1964, 1966, 1967, 1969, 1971 (twice), 1972, 1973, 1974, 1975

—

Copyright © Gerald Durrell, 1958

—

Made and printed in Great Britain by
Cox & Wyman Ltd,
London, Reading and Fakenham
Set in Monotype Bembo

CONTENTS

I. Background for Animals

II. Animals in General

III. Animals in Particular

IV. The Human Animal

INTRODUCTION

DURING the past nine years, between leading expeditions to various parts of the world, catching a multitude of curious creatures, getting married, having malaria, and writing several books, I made a number of broadcasts on different animal subjects for the B.B.C. As a result of these I had many letters asking for copies of the scripts. The simplest way of dealing with this problem was to amass all the various talks in the form of a book, and this I have now done.

That the original talks were at all popular is entirely due to the producers I have had, and in particular Miss Eileen Molony, to whom this book is dedicated. I shall always remember her tact and patience during rehearsals. In a bilious green studio, with the microphone leering at you from the table like a Martian monster, I am never completely at ease. So it was Eileen's unenviable task to counteract the faults in delivery that these nerves produced. I remember with pleasure her voice coming over the intercom. with such remarks as: 'Very good, Gerald, but at the rate you're reading it will be a five-minute talk, not a fifteen-minute one.' Or, 'Try to get a little enthusiasm into your voice there, will you? It sounds as if you hated the animal . . . and try not to sigh when you say your opening sentence . . . you nearly blew the microphone away, and you've no idea how lugubrious it sounded.' Poor Eileen suffered much attempting to teach me the elements of broadcasting, and any success I have achieved in this direction has been entirely due to her guidance. In view of this, it seems rather uncharitable of me to burden her with the dedication of this book, but I know of no other way of thanking her publicly for her help. And anyway, I don't expect her to read it.

Part One

BACKGROUND FOR
ANIMALS

I AM constantly being surprised by the number of people, in different parts of the world, who seem to be quite oblivious to the animal life around them. To them the tropical forests or the savannah or the mountains in which they live, are apparently devoid of life. All they see is a sterile landscape. This was brought home most forcibly to me when I was in Argentina. In Buenos Aires I met a man, an Englishman who had spent his whole life in Argentina, and when he learnt that my wife and I intended to go out into the pampa to look for animals he stared at us in genuine astonishment.

'But, my dear chap, you won't find anything *there*,' he exclaimed.

'Why not?' I inquired, rather puzzled, for he seemed an intelligent person.

'But the pampa is just a lot of grass,' he explained, waving his arms wildly in an attempt to show the extent of the grass, 'nothing, my dear fellow, absolutely nothing but grass punctuated by cows.'

Now, as a rough description of the pampa this is not so very wide of the mark except that life on this vast plain does not consist entirely of cows and gauchos. Standing in the pampa you can turn slowly round and on all sides of you, stretching away to the horizon, the grass lies flat as a billiard-table, broken here and there by the clumps of giant thistles, six or seven feet high, like some extraordinary surrealist candelabra. Under the hot blue sky it does seem to be a dead landscape, but under the shimmering cloak of grass, and in the small forests of dry, brittle thistle-stalks the amount of life is extraordinary. During the hot part of the day, riding on horseback across the thick carpet of grass, or pushing through a giant thistle-forest so that the brittle stems cracked and rattled like fireworks, there was little to be seen except the birds. Every forty or fifty yards there would be burrowing owls, perched straight as guardsmen on a tussock of grass near their holes, regarding you with astonished frosty-cold eyes, and, when you got close, doing a little bobbing dance of anxiety before taking off and wheeling over the grass on silent wings.

Inevitably your progress would be observed and reported on by the watchdogs of the pampa, the black-and-white spurwinged plovers, who would run furtively to and fro, ducking their heads and watching you carefully, eventually taking off and swooping round and round you on piebald wings, screaming 'Tero-tero-tero ... tero ... tero,' the alarm cry that warned everything for miles around of your presence. Once this strident warning had been given, other plovers in the distance would take it up, until it seemed as though the whole pampa rang with their cries. Every living thing was now alert and suspicious. Ahead, from the skeleton of a dead tree, what appeared to be two dead branches would suddenly take wing and soar up into the hot blue sky: chimango hawks with handsome rust-and-white plumage and long slender legs. What you had thought was merely an extra-large tussock of sun-dried grass would suddenly hoist itself up on to long stout legs and speed away across the grass in great loping strides, neck stretched out, dodging and twisting between the thistles, and you realized that your grass tussock had been a rhea, crouching low in the hope that you would pass it by. So, while the plovers were a nuisance in advertising your advance, they helped to panic the other inhabitants of the pampa into showing themselves.

Occasionally you would come across a 'laguana', a small shallow lake fringed with reeds and a few stunted trees. Here there were fat green frogs, but frogs which, if molested, jumped *at* you with open mouth, uttering fearsome gurking noises. In pursuit of the frogs were slender snakes marked in grey, black, and vermilion red, like old school ties, slithering through the grass. In the rushes you would be almost sure to find the nest of a screamer, a bird like a great grey turkey: the youngster crouching in the slight depression in the sun-baked ground, yellow as a buttercup, but keeping absolutely still even when your horse's legs straddled it, while its parents paced frantically about, giving plaintive trumpeting cries of anxiety, intermixed with softer instructions to their chick.

This was the pampa during the day. In the evening, as you rode homeward, the sun was setting in a blaze of coloured clouds, and on the laguanas various ducks were flighting in, arrowing the smooth water with ripples as they landed. Small flocks of spoonbills drifted down like pink clouds to feed in the shallows among snowdrifts of black-necked swans.

As you rode among the thistles and it grew darker you might meet armadillos, hunched and intent, trotting like strange mechanical toys on their nightly scavenging; or perhaps a skunk who would stand, gleaming vividly black and white in the twilight, holding his tail stiffly erect while he stamped his front feet in petulant warning.

This, then, was what I saw of the pampa in the first few days. My friend had lived in Argentina all his life and had never realized that this small world of birds and animals existed. To him the pampa was 'nothing but grass punctuated by cows'. I felt sorry for him.

THE BLACK BUSH

A FRICA is an unfortunate continent in many ways. In Victorian times it acquired the reputation of being the Dark Continent, and even today, when it contains modern cities, railways, macadam roads, cocktail bars, and other necessary adjuncts of civilization, it is still looked upon in the same way. Reputations, whether true or false, die hard, and for some reason a bad reputation dies hardest of all.

Perhaps the most maligned area of the whole continent is the West Coast, so vividly described as the White Man's Grave. It has been depicted in so many stories – quite inaccurately – as a vast, unbroken stretch of impenetrable jungle. If you ever manage to penetrate the twining creepers, the thorns and undergrowth (and it is quite surprising how frequently the impenetrable jungle is penetrated in stories), you find that every bush shakes and quivers with a mass of wild life waiting its chance to leap out at you: leopards with glowing eyes, snakes hissing petulantly, crocodiles in the streams straining every nerve to look more like a log of wood than a log of wood does. If you should escape these dangers there are always the savage native tribes to give the unfortunate traveller the *coup de grâce*. The natives are of two kinds, cannibal and non-cannibal: if they are cannibal, they are always armed with spears; if non-cannibal, they are armed with arrows whose tips drip deadly poison of a kind generally unknown to science.

Now, no one minds giving an author a bit of poetic licence, provided it is recognizable as such. But unfortunately the West Coast of Africa has been libelled to such an extent that anyone who tries to contradict the accepted ideas is branded as a liar who has never been there. It seems to me a great pity that an area of the world where you find nature at its most bizarre, flamboyant and beautiful should be so abused, though I realize that I am a voice crying rather plaintively in the wilderness.

My work has enabled me, one way and another, to see quite a lot of tropical forest, for when you collect live wild animals for a living you have to go out into the so-called impenetrable jungle and look for them. They do not, unfortunately, come to you. It has been brought home to me that in the average tropical forest there is an extraordinary *lack* of wild life: you can walk all day and see nothing more exciting than an odd bird or butterfly. The animals are there, of course, and in rich profusion, but they very wisely avoid you, and in order to see or capture them you have to know exactly where to look. I remember once, after a six months' collecting trip in the forests of the Cameroons, that I showed my collection of about one hundred and fifty different mammals, birds and reptiles to a gentleman who had spent some twenty-five years in that area, and he was astonished that such a variety should have been living, as it were, on his doorstep, in the forest he had considered uninteresting and almost devoid of life.

In the pidgin-English dialect spoken in West Africa, the forest is called the Bush. There are two kinds of Bush: the area that surrounds a village or a town and which is fairly well trodden by hunters and in some places encroached on by farmland. Here the animals are wary and difficult to see. The other type is called the Black Bush, areas miles away from the nearest village, visited by an odd hunter only now and then; and it is here, if you are patient and quiet, that you will see the wild life.

To catch animals, it is no use just scattering your traps wildly about the forest, for, although at first the movements of the animals seem haphazard, you very soon realize that the majority of them have rooted habits, following the same paths year in and year out, appearing in certain districts at certain times when the food supply is abundant, disappearing again when the food fails, always visiting the same places for water. Some of them even have special lavatories which may be some distance away from the place where they spend most of their lives. You may set a trap in the forest and catch nothing in it, then shift that trap three yards to the left or right on to a roadway habitually used by some creature; and thus make your capture at once. Therefore, before you can start on your trapping, you must patiently and carefully investigate the area around you, watching to see which routes are used through the tree-tops or on the forest floor; where supplies of wild fruit are ripening; and which holes are used as bedrooms during the daytime by the nocturnal animals. When I was in West Africa I spent many hours in the Black Bush, watching the forest creatures, studying their habits, so that I would find it more easy to catch and keep them.

I watched one such area over a period of about three weeks. In the Cameroon forests you occasionally find a place where the soil is too shallow to support the roots of the giant trees, and here their place has been taken by the lower growth of shrubs and bushes and long grass which manages to exist on the thin layer of earth covering a grey carapace of rock beneath. I soon found that the edge of one of these natural grassfields, which was about three miles from my camp, was an ideal place to see animal life, for here there were three distinct zones of vegetation: first, the grass itself, five acres in extent, bleached almost white by the sun; then surrounding it a narrow strip of shrubs and bushes thickly entwined with parasitic creepers and hung with the vivid flowers of the wild convolvulus; and finally, behind this zone of low growth, spread the forest proper, the giant tree-trunks a hundred and fifty feet high like massive columns supporting the endless roof of green leaves. By choosing your vantage-point carefully you could get a glimpse of a small section of each of these types of vegetation.

I would leave the camp very early in the morning; even at that hour the sun was fierce. Leaving the camp clearing, I then plunged into the coolness of the forest, into a dim green light that filtered through the multitude of leaves above. Picking my way through the gigantic tree-trunks, I moved across the forest floor, so thickly covered with layer upon layer of dead leaves that it was as soft and springy as a Persian carpet. The only sounds were the incessant zithering of the millions of cicadas, beautiful green-and-silver insects that clung to the bark of the trees, making the air vibrate with their cries, and when you approached too closely zooming away through the forest like miniature aeroplanes, their transparent wings glittering as they flew. Then there would be an occasional plaintive 'whowee' of some small bird which I never managed to identify, but which always accompanied me through the forest, asking questions in its soft liquid tones.

In places there would be a great gap in the roof of leaves above, where some massive branch had perhaps been undermined by insects and damp until it had eventually broken loose and crashed hundreds of feet to the forest floor below, leaving this rent in the forest canopy through which the sunlight sent its golden shafts. In these patches of brilliant light you would find butterflies congregating: large ones with long, narrow, orange-red wings that shone against the darkness of the forest like dozens of candle flames; delicate little white ones like snowflakes would rise in clouds about my feet, then drift slowly back on to the dark leaf-mould, pirouetting as they went. Eventually I reached the banks of a tiny stream which whispered its way through the water-worn boulders, each wearing a cap of green moss and tiny plants. This stream flowed through the forest, through the rim of short growth and out into the grassfield. Just before it reached the edge of the forest, however, the ground sloped and the water flowed over a series of miniature waterfalls, each decorated with clumps of wild begonias whose flowers were a brilliant waxy yellow. Here, at the edge of the forest, the heavy rains had gradually washed the soil from under the massive roots of one of the giant trees which had crashed down and now lay half in the forest and half in the grassfield, a great hollow, gently rotting shell, thickly overgrown with convolvulus, moss, and with battalions of tiny toadstools marching over its peeling bark. This was my hideout, for in one part of the trunk the bark had given way and the hollow interior lay revealed, like a canoe, in which I could sit well hidden by the low growth. When I had made sure that the trunk had no other occupant I would conceal myself and settle down to wait.

For the first hour or so there would be nothing – only the cries of cicadas, an occasional trill from a tree-frog on the banks of the stream, and sometimes a passing butterfly. Within a short time the forest would have forgotten and absorbed you, and after an hour, if you kept still, you would be accepted just as another, if rather ungainly, part of the scenery.

Generally, the first arrivals were the giant plantain-eaters who came to feed on the wild figs which grew round the edge of the grassfield. These huge birds, with long, dangling magpie-like tails, would give notice of their arrival when they were half a mile or so away in the forest, by a series of loud, ringing and joyful cries . . . caroo, coo, coo, coo. They would appear, flying swiftly from the forest with a curious dipping flight, and land in the fig trees, shouting delightedly to each other, flipping their long tails so that their golden-green plumage gleamed iridescently. They would run along the branches in a totally unbirdlike way and leap from one branch to another with great kangaroo jumps, plucking off the wild figs and gulping them down. The next arrivals to the feast would be a troop of Mona monkeys, with their russet-red fur, grey legs, and the two strange, vivid white patches like giant thumbprints on each side of the base of the tail. To hear the monkeys approaching sounded like a sudden wind roaring and rustling through the forest, but if you listened carefully you would hear in the background a peculiar sort of whoop-whoop noise followed by loud and rather drunken honkings, like a fleet of ancient taxicabs caught in a traffic jam. This was the sound of the hornbills, birds who always followed the monkey troops around, feeding not only on the fruit that the monkeys discovered but also on the lizards, tree-frogs, and insects that the movements of the monkeys through the tree-tops disturbed.

On reaching the edge of the forest, the leader of the monkeys would climb to a vantage-point and, uttering suspicious grunts, survey the grassfield in front of him with the greatest care. Behind him the troop, numbering perhaps fifty individuals, would be silent except for the wheezy cry now and again from some tiny baby. Presently, when he was satisfied that the clearing contained nothing, the old male would stalk along a branch slowly and gravely, his tail curled up over his back like a question-mark, and then give a prodigious leap that sent him crashing into the fig-tree leaves. Here he paused again and once more examined the grassfield; then he plucked the first fruit and uttered a series of loud imperative calls: oink, oink, oink. Immediately, the still forest behind him was alive with movement, branches swishing and roaring like giant waves on a beach as the monkeys leapt out of cover and landed in the fig-trees, grunting and squeaking to each other as they plummeted through the air. Many of the female monkeys carried tiny babies which clung under their bellies, and as their mothers jumped you could hear the infant squeaking shrilly, though whether from fear or excitement it was difficult to judge.

The monkeys settled down on the branches to feed on the ripe figs, and presently, with loud swishings and honks of delight, the hornbills discovered their whereabouts and came crashing among the branches in the wild disorderly way in which hornbills always land. Their great round eyes, thickly fringed with heavy eyelashes, stared roguishly and slightly idiotically at the monkeys, while with their enormous and apparently cumbersome beaks they delicately and with great precision plucked the fruit and tossed it carelessly into the air, so that it fell back into their gaping mouths and disappeared. The hornbills were by no means such wasteful feeders as the monkeys, for they would invariably eat each fruit they plucked, whereas the monkeys would take only one bite from a fruit before dropping it to the ground below and moving along the branch to the next delicacy.

The arrival of such rowdy feeding companions was evidently distasteful to the giant plantain-eaters, for they moved off as soon as the monkeys and hornbills arrived. After half an hour or so the ground beneath the fig-trees was littered with half-eaten fruit, and the monkeys then made their way back into the forest, calling oink-oink to each other in a self-satisfied kind of way. The hornbills paused to have just one more fig and then flew excitedly after the monkeys, and as the sound of their wings faded away the next customers for the fig-tree arrived on the scene. They were so small and appeared so suddenly and silently out of the long grass that unless you had field-glasses and kept a careful watch they would come and go without giving a sign of their presence. They were the little striped mice whose homes were amongst the tussocks of grass, the tree-roots, and the boulders along the edge of the forest. Each about the size of a house mouse, with a long and delicately tapering tail, they were clad in sleek, fawny-grey fur which was boldly marked with creamy white stripes from nose to rump. They would drift out from among the grass stalks, moving in little fits and starts, with many long pauses when they sat on their haunches, their tiny pink paws clenched into fists, their noses and whiskers quivering as they tested the wind for enemies. When they froze thus into immobility against the grass stalks, their striped coats, which when they were moving seemed so bright and decorative, acted like an invisible cloak and made them almost disappear.

Having assured themselves that the hornbills had really left (for a hornbill is occasionally partial to a mouse), they set about the serious task of eating the fruit that the monkeys had so lavishly scattered on the ground. Unlike a lot of the other forest mice and rats, these little creatures were of a quarrelsome disposition and would argue over the food, sitting up on their hind legs and abusing each other in thin, reedy squeaks of annoyance. Sometimes two of them would come upon the same fruit and both lay hold of it, one at either end, digging their little pink paws into the leaf mould and tugging frantically in an effort to break the other's grip. If the fig were exceptionally ripe, it generally gave way in the middle so that both mice fell over backwards, each clutching his share of the trophy. They then sat quite peacefully within six inches of one another, each eating his portion. At times, if some sudden noise alarmed them, they all leapt vertically into the air six inches or even more as though suddenly plucked upwards by strings, and on landing they sat quivering and alert until they were sure the danger had passed, when they once more started bickering and fighting over the food.

Once I saw a tragedy enacted among these striped mice as they squabbled amongst the monkeys' left-overs. Suddenly a genet appeared out of the forest. This is perhaps one of the most lithe and beautiful animals to be seen in the forest, with its long sinuously weasel-shaped body and cat-like face, handsome golden fur heavily blotched with a pattern of black spots and long tail banded in black and white. It is not an animal you generally see in the early morning, for its favourite hunting time is in the evenings or at night. I presume this particular one must have had a fruitless night's hunting, and so when morning arrived he was still searching for something to fill his stomach. When he appeared at the edge of the grassfield and saw the striped mice, he flattened close to the ground and then launched himself as smoothly as a skimming stone across the intervening space, and was in amongst the rodents before they knew what was happening. As usual, they all leapt perpendicularly into the air and then fled, looking like portly little business men in their striped suits, rushing through the grass items; but the genet had been too quick and he walked off into the forest, carrying in his mouth two limp little bodies which had so recently been abusing each other as to the sole ownership of a fig and had consequently left it too late to retreat.

Towards midday the whole country fell quiet under the hot rays of the sun, and even the incessant cries of the cicadas seemed to take on a sleepy note. This was siesta time, and very few creatures were to be seen. In the grassfield only the skinks, who loved the sun, emerged to bask on the rocks or to stalk the grasshoppers and locusts. These brilliant lizards, shining and polished as though freshly painted, had skins like mosaic work, made up of hundreds of tiny scales coloured cherry-red, cream and black. They would run swiftly through the grass stalks, their bodies glinting in the sun, so that they looked like some weird firework. Apart from these reptiles, there was practically nothing to be seen until the sun dipped and the day became a trifle cooler, so it was during this period of

inactivity that I used to eat the food I had brought with me and smoke a much-needed cigarette.

Once during my lunch break I witnessed an extraordinary comedy that was performed almost, I felt, for my special benefit. On the tree-trunk where I was sitting, not six feet away, out of a tangle of thick undergrowth, up over the bark of the trunk, there glided slowly and laboriously and very regally a giant land-snail, the size of an apple. I watched it as I ate, fascinated by the way the snail's body glided over the bark, apparently without any muscular effort whatever, and the way its horns with the round, rather surprised eyes on top, twisted this way and that as it picked its route through the miniature landscape of toadstools and moss. Suddenly I realized that as the snail was making its slow and rather vague progress along the trunk it was leaving behind it the usual glistening trail, and this trail was being followed by one of the most ferocious and bloodthirsty animals, for its size, to be found in the West African forest.

The twining convolvulus was thrust aside, and out on to the log strutted a tiny creature only as long as a cigarette, clad in jet-black fur and with a long slender nose that it kept glued to the snail's track, like a miniature black hound. It was one of the forest's shrews, whose courage is incredible and whose appetite is prodigious and insatiable. If anything lives to eat, this forest shrew does. They will even in a moment of peckishness think nothing of eating one another. Chittering to himself, the shrew trotted rapidly after the snail and very soon overtook it. Uttering a high-pitched squeak, it flung itself on that portion of the snail which protruded from the rear of the shell and sank its teeth into it. The snail, finding itself so suddenly and unceremoniously attacked from the rear, did the only possible thing and drew its body rapidly back inside its shell. This movement was performed so swiftly and the muscular contraction of the snail was so strong, that as the tail disappeared inside the shell the shrew's face was banged against it and his grip was broken. The shell, having now nothing to balance it, fell on its side, and the shrew, screaming with frustration, rushed forward and plunged his head into the interior, in an effort to retrieve the retreating mollusc. However, the snail was prepared for this attack and as soon as the shrew's head was pushed into the opening of the shell it was greeted by a sudden fountain of greenish-white froth that bubbled out and enveloped nose and head. The shrew leapt back with surprise, knocking against the shell as it did so. The snail teetered for a moment and then rolled sideways and dropped into the undergrowth beneath the log. The shrew meanwhile was sitting on its hind legs, almost incoherent with rage, sneezing violently and trying to wipe the froth from its face with its paws. The whole thing was so ludicrous that I started to laugh, and the shrew, casting a hasty and frightened glance in my direction, leapt down into the undergrowth and hurried away. It was not often during the forest's siesta-time that I could enjoy such a scene as this.

At mid-afternoon, when the heat had lessened, the life of the forest would start again. There were new visitors to the fig-trees, in particular the squirrels. There was one pair who obviously believed in combining business with pleasure, for they ran and leapt among the fig-tree branches, playing hide-and-seek and leap-frog, courting each other in this way, and occasionally breaking off their wild and exuberant activities to sit very quietly and solemnly, their tails draped over their backs, eating figs. As the shadows grew longer, you might, if you were lucky, see a duiker coming down to drink at the stream. These small antelopes, clad in shining russet coats, with their long, pencil-slim legs, would pick their way slowly and suspiciously through the forest trees, pausing frequently while their large liquid eyes searched the path ahead, and their ears twitched backwards and forwards, picking up the sounds of the forest. As they drifted their way without a sound through the lush plants bordering the stream, they would generally disturb some of the curious aquatic mice who were feeding there. These little grey rodents have long, rather stupid-looking faces, big semi-transparent ears shaped like a mule's, and long hind legs on which they would at times hop like miniature kangaroos. At this hour of the evening it was their habit to wade through the shallow water, combing the water-weed with their slender front paws and picking out tiny water-insects, baby crabs and water-snails. At this time rats of another type would also come out, and these were probably the most fussy, pompous and endearing of the rodents. They were clad all over in greenish fur, with the exception of their noses and their behinds which were, rather incongruously, a bright foxy red, and made them look rather as though they were wearing red running shorts and masks. Their favourite hunting-ground was in the soft leaf-mould between the towering buttress roots of the great trees. Here they would waddle about, squeaking to each other, turning over leaves and bits of rock and twigs for the insects which were concealed underneath. Occasionally they would stop and hold conversations, sitting on their hind legs, facing each other, their whiskers trembling as they chittered and squeaked very rapidly and in a complaining sort of tone as though commiserating with one another on the lack of food in that particular part of the forest. There were times when, sniffing about in certain patches, they became terribly excited, squeaking loudly and digging, like terriers, down into the soft leaf-mould. Eventually they would triumphantly unearth a big chocolate-coloured beetle, almost as large as themselves. These insects were horny and very strong, and it took the rats a good deal of effort to subdue them. They would turn them on to their backs and then rapidly nip off the spiky, kicking legs. Once they had immobilized their prey, a couple of quick bites and the beetle was dead. Then the little rat would sit up on its hind legs, clasping the body of the beetle in both hands and proceed to eat it, as though it were a stick of rock, with loud scrunchings and occasional muffled squeaks of delight.

By now, although still light in the grassfield, it was gloomy and difficult to see in the forest itself. You might, if you were fortunate, catch a glimpse of some of the nocturnal animals venturing out on their hunting: perhaps a brush-tailed porcupine would trot past, portly and serious, his spines rustling through the leaves as he hurried on his way. Now the fig-trees would once again become the focal-point, as these nocturnal animals appeared. The galagos, or bush-babies, would materialize magically, like fairies, and sit among the branches, peering about them with their great saucer-shaped eyes, and their little incredibly human-looking hands held up in horror, like a flock of pixies who had just discovered that the world was a sinful place. They would feed on the figs and sometimes take prodigious leaps through the branches in pursuit of a passing moth, while overhead, across a sky already flushed with sunset colours, pairs of grey parrots flew into the forest to roost, whistling and cooing to each other, shrilly, so that the forest echoed. Far away in the distance a chorus of hoots suddenly rose, screams and wild bursts of maniacal laughter, the hair-raising noise of a troop of chimpanzees going to bed. The galagos would now have disappeared as quickly and as silently as they had come, and through the darkening sky the fruit-bats would appear in great tattered clouds, flapping down, giving their ringing cries, diving into the trees to fight and flutter round the remains of the fruit, so that the sound of their wings was like a hundred wet umbrellas being shaken amongst the trees. There would be one more shrill and hysterical outburst from the chimpanzees, and then the forest was completely dark, but still alive and vibrating with a million little rustles, squeaks, patters, and grunts, as the night creatures took over.

I rose to my feet, cramped and stiff, and stumbled off through the forest, the glow of my torch seeming pathetically frail and tiny among the great silent tree-trunks. This, then, was the tropical forest that I had read about as savage, dangerous and unpleasant. To me it seemed a beautiful and incredible world made up of a million tiny lives, plants and animals, each different and yet dependent on the other, like the many parts of a gigantic jigsaw puzzle. It seemed to me such a pity that people should still cling to their old ideas of the unpleasantness of the jungle when here was a world of magical beauty waiting to be explored, observed and understood.

LILY-TROTTER LAKE

BRITISH GUIANA, lying in the northern part of South America, is probably one of the most beautiful places in the world, with its thick tropical forest, its rolling savannah land, its jagged mountain ranges and giant foaming waterfalls. To me, however, one of the most lovely parts of Guiana is the creek lands. This is a strip of coastal territory that runs from Georgetown to the Venezuelan border; here a thousand forest rivers and streams have made their way down towards the sea, and on reaching the flat land have spread out into a million creeks and tiny waterways that glimmer and glitter like a flood of quicksilver. The lushness and variety of the vegetation is extraordinary, and its beauty has turned the place into an incredible fairyland. In 1950 I was in British Guiana collecting wild animals for zoos in England, and during my six months there I visited the savannah lands to the north, the tropical forest and, of course, the creek lands, in pursuit of the strange creatures living there.

I had chosen a tiny Amerindian village near a place called Santa Rosa as my headquarters in the creek lands, and to reach it required a two-day journey. First by launch down the Essequibo River and then through the wider creeks until we reached the place where the launch could go no farther, for the water was too shallow and too choked with vegetation. Here we took to dug-out canoes, paddled by the quiet and charming Indians who were our hosts, and leaving the broad main creek we plunged into a maze of tiny waterways on one of the most beautiful journeys I can remember.

Some of the creeks along which we travelled were only about ten feet wide, and the surface of the water was completely hidden under a thick layer of great creamy water-lilies, their petals delicately tinted pink, and a small fern-like water-plant that raised, just above the surface of the water, on a slender stem, a tiny magenta flower. The banks of the creek were thickly covered with undergrowth and great trees, gnarled and bent, leant over the waters to form a tunnel; their branches were festooned with long streamers of greenish-grey Spanish moss and clumps of bright pink-and-yellow orchids. With the water so thickly covered with vegetation, you had the impression, when sitting in the bows of the canoe, that you were travelling smoothly and silently over a flower-studded green lawn that undulated gently in the wake of your craft. Great black woodpeckers, with scarlet crests and whitish beaks, cackled loudly as they flipped from tree to tree, hammering away at the rotten bark, and from the reeds and plants along the edges of the creek there would be a sudden explosion of colour as we disturbed a marsh bird which flew vertically into the air, with its hunting-pink breast flashing like a sudden light in the sky.

The village, I discovered, was situated on an area of high ground which was virtually an island, for it was completely surrounded by a chess-board of creeks. The little native hut that was to be my headquarters was some distance away from the village and placed in the most lovely surroundings. On the edge of a tiny valley an acre or so in extent, it was perched amongst some great trees which stood round it like a group of very old men with long grey beards of Spanish moss. During the winter rains the surrounding creeks had overflowed so that the valley was now drowned under some six feet of water out of which stuck a number of large trees, their reflections shimmering in the sherry-coloured water. The rim of the valley had grown a fringe of reeds and great patches of lilies. Sitting in the doorway of the hut, one had a perfect view of this miniature lake and its surroundings, and it was sitting here quietly in the early morning or evening that I discovered what a wealth of animal life inhabited this tiny patch of water and its surrounding frame of undergrowth.

In the evening, for example, a crab-eating raccoon would come down to drink. They are strange-looking animals, about the size of a small dog, with bushy tails ringed in black and white, large, flat, pink paws, the grey of their body-fur relieved only by a mask of black across the eyes, which gives the creature a rather ludicrous appearance. These animals walk in a curious humpbacked manner with their feet turned out, shuffling along in this awkward fashion like someone afflicted with chilblains. The raccoon came down to the water's edge and, having stared at his reflection dismally for a minute or so, drank a little and then with a pessimistic air shuffled slowly round the outer rim of the valley in search of food. In patches of shallow water he would wade in a little way and, squatting on his haunches, feel about in the dark water with the long fingers of his front paws, patting and touching and running them through the mud, and he would suddenly extract something with a look of pleased surprise and carry it to the bank to be eaten. The trophy was always carried clasped delicately between his front paws and dealt with when he arrived on dry land. If it was a frog, he would hold it down and with one quick snap decapitate it. If, however, as was often the case, it was one of the large freshwater crabs, he would hurry shorewards as quickly as possible, and on reaching land flick the crab away from him. The crab would recover its poise and menace him with open pincers, and the raccoon would then deal with it in a very novel and practical way. A crab is very easily discomfited, and if you keep tapping at it and it finds that every grab it makes at you with its pincers misses the mark, it will eventually fold itself up and sulk, refusing to participate any more in such a one-sided contest. So the raccoon simply followed the crab around, tapping him on his carapace with his long fingers and whipping them out of the way every time the pincers came within grabbing distance. After five minutes or so of this the frustrated crab would fold up and just squat. The raccoon, who till then had resembled a dear old lady playing with a Pekinese, would straighten up and become businesslike, and, leaning forward, with one quick snap would cut the unfortunate crab almost in two.

Along one side of the valley some previous Indian owner of the hut had planted a few mango and guava trees, and while I was there the fruit ripened and attracted a great number of creatures. The tree-porcupines were generally the first on the scene. They lumbered out of the undergrowth, looking like portly and slightly inebriated old men, their great bulbous noses whiffling to and fro, while their tiny and rather sad little eyes, that always seemed full of unshed tears, peered about them hopefully. They climbed up into the mango-trees very skilfully, winding their long, prehensile tails round the branches to prevent themselves from falling, their black-and-white spines rattling among the leaves. They then made their way along to a comfortable spot on a branch, anchored themselves firmly with a couple of twists of the tail, then sat up on their hind legs, and plucked off a fruit. Holding it in their front paws, they turned it round and round while their large buck teeth got to work on the flesh. When they had finished a mango they sometimes began playing a rather odd game with the big seed. Sitting there they looked round in a vague and rather helpless manner while juggling the seed from paw to paw as though not quite certain what to do with it, and occasionally pretending to drop it and recovering it at the last moment. After about five minutes of this they tossed the seed down to the ground below and shuffled about the tree in search of more fruit.

Sometimes when one porcupine met another face to face on a branch, they both anchored themselves with their tails, sat up on their hind legs and indulged in the most ridiculous boxing-match, ducking, and slapping with their front paws, feinting and lunging, giving left hooks, uppercuts and body blows, but never once making contact. Throughout this performance (which lasted perhaps for a quarter of an hour) their expression never changed from one of bewildered and benign interest. Then, as though prompted by an invisible signal, they went down on all fours and scrambled away to different parts of the tree. I could never discover the purpose of these boxing bouts nor identify the winner, but they afforded me an immense amount of amusement.

Another fascinating creature that used to come to the fruit trees was the douroucouli. These curious little monkeys, with long tails, delicate, almost squirrel-like bodies and enormous owl-like eyes, are the only nocturnal species of monkey in the world. They arrived in small troops of seven or eight and, though they made no noise as they jumped into the fruit trees, you could soon tell they were there by the long and complicated conversation they held while they fed. They had the biggest range of noises I have ever heard from a monkey, or for that matter from any animal of similar size. First they could produce a loud purring bark, a very powerful vibrating cry which they used as a warning; when they delivered it their throats would swell up to the size of a small apple with the effort. Then, to converse with one another, they would use shrill squeaks, grunts, a mewing noise not unlike a cat's and a series of liquid, bubbling sounds quite different from anything else I have ever heard. Sometimes one of them in an excess of affection would drape his arm over a companion's shoulder and they would sit side by side, arms round each other, bubbling away, peering earnestly into each other's faces. They were the only monkeys I know that would on the slightest provocation give one another the most passionate human kisses, mouth to mouth, arms round each other, tails entwined.

Naturally these animals made only sporadic appearances; there were, however, two creatures which were in constant evidence in the waters of the drowned valley. One was a young cayman, the South American alligator, about four feet long. He was a very handsome reptile with black-and-white skin as knobbly and convoluted as a walnut, a dragon's fringe on his tail, and large eyes of golden-green flecked with amber. He was the only cayman to live in this little stretch of water. I could never understand why no others had joined him, for the creeks and waterways, only a hundred feet or so away, were alive with them. None the less this little cayman lived in solitary state in the pool outside my hut and spent the day swimming round and round with a rather proprietory air. The other creature always to be seen was a jacana, probably one of the strangest birds in South America. In size and appearance it is not unlike the English moorhen, but its neat body is perched on long slender legs which end in a bunch of enormously elongated toes. It is with the aid of these long toes and the even distribution of weight they give that the jacana manages to walk across water, using the water-lily leaves and other water-plants as its pathways. It has thus earned its name of lily-trotter.

The jacana disliked the cayman, while the cayman had formed the impression that Nature had placed the jacana in his pool to add a little variety to his diet. He was, however, a young and inexperienced reptile, and at first his attempts to stalk and capture the bird were ridiculously obvious. The jacana would come mincing out of the undergrowth, where it used to spend much of its time, and walk out across the water, stepping delicately from one lily leaf to the next, its long toes spreading out like spiders and the leaves dipping gently under its weight. The cayman, on spotting it, immediately submerged until only his eyes showed above water. No ripple disturbed the surface, yet his head seemed to glide along until he got nearer and nearer to the bird. The jacana, always pecking busily among the water-plants in search of worms and snails and tiny fish, rarely noticed the cayman's approach and would probably have fallen an easy victim if it had not been for one thing. As soon as the cayman was within ten or twelve feet he would become so excited that instead of submerging and taking the bird from underneath he would suddenly start to wag his tail vigorously and shoot along the surface of the water like a speedboat, making such large splashes that not even the most dim-witted bird could have been taken unawares; and the jacana would fly up into the air with a shrill cry of alarm, wildly flapping its buttercup-yellow wings.

For a long time it did not occur to me to wonder why the bird should spend a greater part of the day in the reed-bed at one end of the lake. But on investigating this patch of reed I soon discovered the reason, for there on the boggy ground I found a mat neatly made of weed on which lay four round creamy eggs heavily blotched with chocolate and silver. The bird must have been sitting for some time, for only a couple of days later I found the nest empty and a few hours after that saw the jacana leading out her brood for its first walk into the world.

She emerged from the reed-bed, trotted out on to the lily leaves, then paused and looked back. Out of the reeds her four babies appeared, with the look of outsize bumble-bees, in their golden-and-black fluff, while their long slender legs and toes seemed as fragile as spider-webs. They walked in single file behind their mother, always a lily leaf behind, and they waited patiently for their mother to test everything before moving forward. They could all cluster on one of the great plate-like leaves, and they were so tiny and light that it scarcely dipped beneath their weight. Once the cayman had seen them, of course, he redoubled his efforts, but the jacana was a very careful mother. She kept her brood near the edge of the lake, and if the cayman showed any signs of approaching, the babies immediately dived off the lily leaves and vanished into the water, to reappear mysteriously on dry land a moment later.

The cayman tried every method he knew, drifting as close as possible without giving a sign, concealing himself by plunging under a mat of water-weeds and then surfacing so that the weeds almost covered eyes and nose. There he lay patiently, sometimes even moving very close inshore, presumably in the hope of catching the jacanas before they ventured out too far. For a week he tried each of these methods in turn, and only once did he come anywhere near success. On this particular day he had spent the hot noon hours lying, fully visible, in the very centre of the lake, revolving slowly round and round so that he could keep an eye on what was happening on the shore. In the late afternoon he drifted over to the fringe of lilies and weeds and managed to catch a small frog that had been sunning itself in the centre of a lily. Fortified by this, he swam over to a floating raft of green weeds, studded with tiny flowers, and dived right under it. It was only after half an hour of fruitless search in other parts of the little lake that I realized he must be concealed under the weeds. I trained my field-glasses on them, and although the entire patch was no larger than a door, it took me at least ten minutes to spot him. He was almost exactly in the centre and as he had risen to the surface a frond of weed had become draped between his eyes; on the top of this was a small cluster of pink flowers. He looked somewhat roguish with this weed on his head, as though he were wearing a vivid Easter bonnet, but it served to conceal him remarkably well. Another half an hour passed before the jacanas appeared and the drama began.

The mother, as usual, emerged suddenly from the reed-bed, and stepping daintily on to the lily leaves paused and called her brood, who pattered out after her like a row of quaint clock-work toys and then stood patiently clustered on a lily leaf, awaiting instructions. Slowly the mother led them out into the lake, feeding as they went. She would poise herself on one leaf and, bending over, catch another in her beak, which she would pull and twist until it was sufficiently out of the water to expose the underside. A host of tiny worms and leeches, snails and small crustaceans, generally clung to it. The babies clustered round and pecked vigorously, picking off all this small fry until the underside of the leaf was clean, whereupon they all moved off to another.

Quite early in the proceedings I realized that the female was leading her brood straight towards the patch of weeds beneath which the cayman was hiding, and I remembered then that this particular area was one of her favourite hunting-grounds. I had watched her standing on the lily pads, pulling out the delicate, fern-like weed in large tangled pieces and draping it across a convenient lily flower so that her babies could work over it for the mass of microscopic life it contained. I felt sure that, having successfully managed to evade the cayman so far, she would notice him on this occasion, but although she paused frequently to look about her, she continued to lead her brood towards the reptile's hiding-place.

I was now in a predicament. I was determined that the cayman was not going to eat either the female jacana or her brood if I could help it, but I was not quite sure what to do. The bird was quite used to human noises and took no notice of them whatever, so there was no point in clapping my hands. Nor was there any way of getting close to her, for this scene was being enacted on the other side of the lake, and it would have taken me ten minutes to work my way round, by which time it would be too late, for already she was within twenty feet of the cayman. It was useless to shout, too far to throw stones, so I could only sit there with my eyes glued to my field-glasses, swearing that if the cayman so much as touched a feather of my jacana family I would hunt him out and slaughter him. And then I suddenly remembered the shotgun.

It was, of course, too far for me to shoot at the cayman: the shot would have spread out so much by the time it reached the other side of the lake that only a few pellets would hit him, whereas I might easily kill the birds I was trying to protect. It occurred to me, however, that as far as I knew the jacana had never heard a gun, and a shot fired into the air might therefore frighten her into taking her brood to safety. I dashed into the hut and found the gun, and then spent an agonizing minute or two trying to remember where I had put the cartridges. At last I had it loaded and hurried out to my vantage-point again. Holding the gun under my arm, its barrels pointing into the soft earth at my feet, I held the field-glasses up in my other hand and peered across the lake to see if I was in time.

The jacana had just reached the edge of the lilies nearest the weed patch. Her babies were clustered on a leaf just behind and to one side of her. As I looked she bent forward, grabbed a large trailing section of weed and pulled it on to the lily leaves, and at that moment the cayman, only about four feet away, rose suddenly from his nest of flowers and weeds and, still wearing his ridiculous bonnet, charged forward. At the same moment I let off both barrels of the shotgun, and the roar echoed round the lake.

Whether it was my action that saved the jacana or her own quick-wittedness I do not know, but she rose from the leaf with extraordinary speed just as the cayman's jaws closed and cut the leaf in half. She swooped over his head, he leapt half out of the water in an effort to grab her (I could hear the clop of his jaws) and she flew off unhurt but screaming wildly.

The attack had been so sudden that she had apparently given no orders to her brood, who had meanwhile been crouching on the lily leaf. Now, hearing her call, they were galvanized into action, and as they dived overboard the cayman swept towards them. By the time he reached the spot they were under water, so he dived too and gradually the ripples died away and the surface of the water became calm. I watched anxiously while the female jacana, calling in agitation, flew round and round the lake. Presently she disappeared into the reed-bed and I saw her no more that day. Nor did I see the cayman for that matter. I had a horrible feeling that he had succeeded in catching all those tiny bundles of fluff as they swam desperately under water, and I spent the evening planning revenge.

The next morning I went round to the reed-bed, and there to my delight I found the jacana, and with her three rather subdued-looking babies. I searched for the fourth one, but as he was nowhere to be seen it was obvious that the cayman had been at any rate partially successful. To my consternation the jacana, instead of being frightened off by her experience of the previous day, proceeded once more to lead her brood out to the water-lilies, and for the rest of the day I watched her with my heart in my mouth. Though there was no sign of the cayman, I spent several nerve wracking hours, and by evening I decided I could stand it no longer. I went to the village and borrowed a tiny canoe which two Indians kindly carried down to the little lake for me. As soon as it was dark I armed myself with a powerful torch and a long stick with a slip-knot of rope on the end, and set off on my search for the cayman. Though the lake was so small, an hour had passed before I spotted him, lying on the surface near some lilies. As the torch-beam caught him, his great eyes gleamed like rubies. With infinite caution I edged closer and closer until I could gently lower the noose and pull it carefully over his head, while he lay there quietly, blinded or mesmerized by the light. Then I jerked the noose tight and hauled his thrashing and wriggling body on board, his jaws snapping and his throat swelling as he gave vent to loud harsh barks of rage. I tied him up in a sack and the next day took him five miles deep into the creeks and let him go. He never managed to find his way back, and for the rest of my stay in the little hut by the drowned valley I could sit and enjoy the sight of my lily-trotter family pottering happily over the lake in search of food, without suffering any anxiety every time a breeze ruffled the surface of the rich tawny water.

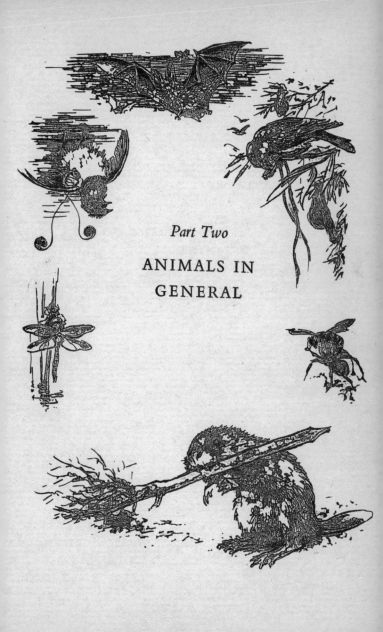

Part Two

ANIMALS IN
GENERAL

THE way animals behave, the way they cope with the problems of existence, has always been a source of fascination to me. In the following talks I tried to show some of the astonishing methods they use to obtain a mate, to defend themselves or to build their homes.

An ugly or horrifying animal – like an ugly or horrifying human being – is never completely devoid of certain attractive qualities. And one of the most disarming things about the animal world is the sudden encounter with what appeared to be a very dull and nasty beast behaving in a charming and captivating way: an earwig squatting like a hen over her nest of eggs, and carefully gathering them all together again if you are unkind enough to scatter them; a spider who, having tickled his lady-love into a trance, takes the precaution of tying her down with silk threads so that she will not suddenly wake up and devour him after the mating; the sea-otter that carefully ties itself to a bed of seaweed so that it may sleep without fear of being carried too far away by the tides and currents.

I remember once, when I was quite young, sitting on the banks of a small sluggish stream in Greece. Suddenly, out of the water crawled an insect that looked as if it had just arrived from outer space. It made its laborious way up the stalk of a bullrush. It had great bulbous eyes, a carunculated body supported on spidery legs, and slung across its chest was a curious contraption, carefully folded, that looked as though it might be some sort of Martian aqualung. The insect made its way carefully up the bullrush while the hot sun dried the water off its ugly body. Then it paused and appeared to go into a trance.

I was fascinated and yet interested by its repulsive appearance, for in those days my interest in natural history was only equalled by my lack of knowledge, and I did not recognize it for what it was. Suddenly I noticed that the creature, now thoroughly sun-dried and as brown as a nut, had split right down its back and, as I watched, it seemed as though the animal inside was struggling to get out. As the minutes passed, the struggles grew stronger and the split grew larger, and presently the creature inside hauled itself free of its ugly skin and crawled feebly on to the rush stalk, and I saw it was a dragonfly. Its wings were still wet and crumpled from this strange birth, and its body soft, but, as I watched, the sun did its work and the wings dried stiff and straight, as fragile as snowflakes and as intricate as a cathedral window. The body also stiffened, and changed to a brilliant sky-blue. The dragonfly whirred its wings a couple of times, making them shimmer in the sun, and then flew unsteadily away, leaving behind, still clinging to the stem, the unpleasant shell of its former self.

I had never seen such a transformation before, and as I gazed with amazement at the unattractive husk which had housed the beautiful shining insect, I made a vow that never again would I judge an animal by its appearance.

ANIMAL COURTSHIPS

Most animals take their courtship very seriously, and through the ages some of them have evolved fascinating ways of attracting the female of their choice. They have grown a bewildering mass of feathers, horns, spikes, and dewlaps, and have developed an astonishing variety of colours, patterns, and scents, all for the purpose of obtaining a mate. Not content with this, they will sometimes bring the female a gift, or construct a flower show for her, or intrigue her with an acrobatic display, or a dance, or a song. When the animals are courting they put their heart and soul into it, and are even, if necessary, ready to die.

The Elizabethan lovers of the animal world are, of course, the birds: they dress themselves magnificently, they dance and posture and they are prepared at a moment's notice to sing a madrigal or fight a duel to the death.

The most famous are the birds of paradise, for not only do they possess some of the most gorgeous courting dresses in the world but they show them off so well.

Take, for example, the king bird of paradise. I was once lucky enough to see one of these birds displaying in a Brazilian zoo. Here, in a huge outdoor aviary full of tropical plants and trees, three king birds of paradise were living – two females and a male. The male is about the size of a blackbird, with a velvety orange head contrasting vividly with a snow-white breast and a brilliant scarlet back, the feathers having such a sheen on them that they seem polished. The beak is yellow and the legs are a beautiful cobalt blue. The feathers along the side of its body – since it was the breeding season – were long, and the middle pair of tail feathers were produced in long slender stalks about ten inches in length. The feather was tightly coiled like a watch spring, so that at the end of each of these wire-like feathers shone a disc of emerald green. In the sunlight he gleamed and glittered with every movement, and the slender tail-wire trembled and the green disc shook and scintillated in the sun. He was sitting on a long bare branch, and the two females were squatting in a bush close by, watching him. Suddenly he puffed himself out a little and gave a curious cry midway between a whine and a yap. He was silent for a minute as if watching the effect of this sound on the females; but they continued to sit there, observing him unemotionally. He bobbed once or twice on the branch, to fix their attention perhaps, then raised his wings above his back and flapped them wildly, just as if he were about to take off on a triumphant flight. He spread them out wide and ducked forward, so that his head was hidden by the feathers. Then he raised them again, flapped vigorously once more, and wheeled round so that the two females should be dazzled by his beautiful snow-white breast. He gave a lovely liquid warbling cry, while the long side-plumes on his body suddenly burst out, like a beautiful fountain of ash-grey, buff and emerald-green that quivered delicately in time to his song. He raised his short tail and pressed it closely to his back, so that the two long tail-wires curved over his head and on each side of his yellow beak hung the two emerald-green discs. He swayed his body gently to and fro; the discs swung like pendulums and gave the odd impression that he was juggling with them. He lifted and lowered his head, singing with all his might, while the green discs gyrated to and fro.

The females seemed completely oblivious. They sat there regarding him with the mild interest of a couple of housewives at an expensive mannequin parade, who, though they admired the gowns, realized they have no hope of purchasing them.

Then the male, as if in a last desperate effort to work his audience into some show of enthusiasm, suddenly swung right round on the perch and showed his beautiful scarlet back to them, lowered himself to a crouch and opened wide his beak, revealing the interior of his mouth which was a rich apple-green in colour and as glossy as though it had just been painted. He stood like this, singing with open mouth, and then gradually, as his song died away, his gorgeous plumes ceased to twitch and tremble and fell against his body. He stood upright on the branch for a moment and regarded the females. They stared back at him with the expectant air of people who, having watched a conjurer performing one good trick, are waiting for the next. The male gave a few slight chirrups and then burst into song again and suddenly let himself drop, so that he hung beneath the branch. Still singing, he spread his wings wide and then walked to and fro upside down. This acrobatic display seemed to intrigue one of the females for the first time, for she cocked her head on one side in a gesture of inquiry. I could not for the life of me see how they could remain so unimpressed, for I was dazzled and captivated by the male's song and colouring. Having walked backwards and forwards for a minute or so, he closed his wings tightly and let his body dangle straight down, swaying softly from side to side, singing passionately all the while. He looked like some weird crimson fruit attached to the bough by the blue stalks of his legs, stirring gently in a breeze.

At this point, one of the females grew bored and flew off to another part of the aviary. The remaining one, however, with head cocked to one side, was peering closely at the male. With a quick flap of his wings he regained his upright position on the perch, looking a trifle smug, I thought, as well he might. Now I waited excitedly to see what would happen next. The male was standing stock-still, letting his feathers shimmer in the sunlight, and the female was becoming decidedly excited. I felt sure that she had succumbed to his fantastic courtship, which was as sudden and as beautiful as a burst of highly coloured fireworks. Sure enough, the female took wing. Now, I thought, she was going up to congratulate him on his performance and they would start married life together at once. But to my astonishment she merely flew on to the branch where the male sat, picked up a small beetle, which was wandering aimlessly across the bark, and with a satisfied chuck flew off to the other end of the aviary with it. The male puffed himself out and started to preen in a resigned sort of fashion, and I decided that the females must be especially hard-hearted, or especially inartistic, to have been able to resist such an exhibition. I felt very sorry for the male that his magnificent courtship should go unrewarded. However my sympathy was wasted, for with a squeak of triumph he had discovered another beetle, which he was happily banging on the branch, He obviously did not mind in the least being turned down.

Not all birds are such good dancers as the birds of paradise, nor are they so well dressed, but they have compensated for this by the charming originality of their approach to the opposite sex. Take, for example, the bower-birds. They have, in my opinion, one of the most delightful courtship methods in the world. The satin bower-bird, for instance, is not particularly impressive to look at: about the size of a thrush, he is clad in dark blue feathers that have a metallic glint when the light catches them. He looks, quite frankly, as if he is wearing an old and shiny suit of blue serge, and you would think that his chances of inducing the female to overlook the poverty of his wardrobe were non-existent. But he contrives it by an extremely cunning device: he builds a bower.

Once again it was in a zoo that I was lucky enough to see a satin bower-bird building his temple of love. He had chosen two large tussocks of grass in the middle of his aviary and had carefully cleared a large circular patch all round and a channel between them. Then he had carried twigs, pieces of string and straw, and had woven them into the grass, so that the finished building was like a tunnel. It was at this stage that I first noticed what he was doing, and by then, having built his little week-end cottage, he was in the process of decorating it. Two empty snail-shells were the first items, and they were followed by the silver paper from a packet of cigarettes, a piece of wool that he had picked up, six coloured pebbles and a bit of string with a blob of sealing wax on it. Feeling that he might like some more decoration, I brought him some strands of coloured wool, a few multi-coloured sea-shells and some bus tickets.

He was delighted; he came down to the wire to take them carefully from my fingers, and then hopped back to his bower to arrange them. He would stand staring at the decorations for a minute and then hop forward and move a bus-ticket or a strand of wool into what he considered a more artistic position. When the bower was finished it really looked very charming and decorative, and he stood in front of it preening himself and stretched out one wing at a time as if indicating his handwork with pride. Then he dodged to and fro through his little tunnel, rearranged a couple of sea-shells, and started posturing again with one wing outspread. He had really worked hard on his bower, and I felt sorry for him, for the whole effort was in vain: his mate had apparently died some time previously and he now shared the aviary with a few squawking common finches that took no interest whatsoever in his architectural abilities or in his display of household treasures.

In the wild state, the satin bower-bird is one of the few birds that uses a tool, for he will sometimes paint the twigs used in the construction of his bower with highly coloured berries and moist charcoal, using a piece of some fibrous material as a brush. Unfortunately, by the time I had remembered this and had made plans to provide my bower-bird with a pot of blue paint and a piece of old rope – the bower-birds are particularly fond of blue – he had lost interest in his bower and not even the presentation of a complete set of cigarette cards, depicting soldiers' uniforms through the ages, could arouse his enthusiasm again.

Another species of bower-bird build an even more impressive structure, four to six feet high, by piling sticks round two trees and then roofing it over with creepers. The inside is carefully laid with moss, and the outside, for this bower-bird is plainly a man of the world with expensive tastes, is decorated with orchids. In front of the bower he constructs a little bed of green moss on which he places all the brightly coloured flowers and berries he can find; being a fastidious bird, he renews these every day, taking the withered decorations and piling them carefully out of sight behind the bower.

Among the mammals, of course, you do not come across quite such displays as among the birds. On the whole, mammals seem to have a more down-to-earth, even modern attitude, towards their love problems.

I was able to watch the courtship of two tigers when I worked at Whipsnade Zoo. The female was a timid, servile creature, cringing at the slightest snarl from her mate until the day she came into season. Then she changed suddenly into a slinking, dangerous creature, fully aware of her attraction but biding her time. By the end of the morning the male was following her round, belly-crawling and abject, while on his nose were several deep, bloody grooves caused by slashing backhands from his mate. Every time he forgot himself and approached too closely he got one of these backhand swipes across the nose. If he seemed at all offended by this treatment

and lay down under a bush, the female would approach him, purring loudly, and rub herself against him until he got up and followed her again, moving closer and closer until he received another blow on the nose for his pains.

Eventually the female led him down into a little dell where the grass was long, and there she lay down and purred to herself, with her green eyes half-closed. The tip of her tail, like a big black-and-white bumble-bee, twitched to and fro in the grass, and the poor besotted male chased it from side to side, like a kitten, slapping it gently with his great paws. At last the female tired of her vamping; she crouched lower in the grass and gave a curious purring cry. The male, rumbling in his throat, moved towards her. She cried again, and raised her head, while the male gently bit along the line of her arched neck, a gentle nibble with his great teeth. Then the female cried again, a self-satisfied purr, and the two great striped bodies seemed to melt together in the green grass.

Not all mammals are so decorative and highly coloured as the tiger, but they generally compensate by being brawny. They therefore have to rely on cave-man tactics for obtaining their mates. Take, for example, the hippopotamus. To see one of these great chubby beasts lying in the water, staring at you with a sort of benign innocence out of bulbous eyes, sighing occasionally in a smug and lethargic manner, would scarcely lead you to believe that they could be roused to bursts of terrible savagery when it came to choosing a mate. If you have ever seen a hippo yawning, displaying on each side of its mouth four great curved razor-sharp tusks (hidden among which two more point outwards like a couple of ivory spikes) you will realize what damage they could do.

When I was collecting animals in West Africa we once camped on the banks of a river in which lived a hippo herd of moderate size. They seemed a placid and happy group, and every time we went up or down the stream by canoe they would follow us a short distance, swimming nearer and nearer, wiggling their ears and occasionally snorting up clouds of spray, as they watched us with interest. As far as I could make out, the herd consisted of four females, a large elderly male and a young male. One of the females had a medium-sized baby with her which, though already large and fat, was still occasionally carried on her back. They seemed, as I say, a very happy family group. But one night, just as it was growing dark, they launched into a series of roars and brays which sounded like a choir of demented donkeys. These were interspersed with moments of silence broken only by a snort or a splash, but as it grew darker the noise became worse, until, eventually realizing I would be unlikely to get any sleep, I decided to go down and see what was happening. Taking a canoe, I paddled down to the curve of the river a couple of hundred yards away, where the brown water had carved a deep pool out of the bank and thrown up a great half-moon of glittering white sand. I knew the hippos liked to spend the day here, and it was from this direction that all the noise was coming. I knew something was wrong, for usually by this time each evening they had hauled their fat bodies out of the water and trekked along the bank to raid some unfortunate native's plantation, but here they were in the pool, long past the beginning of their feeding-time. I landed on the sandbank and walked along to a spot which gave me a good view. There was no reason for me to worry about noise: the terrible roars and bellows and splashes coming from the pool were quite sufficient to cover the scrunch of my footsteps.

At first I could see nothing but an occasional flash of white where the hippos' bodies thrashed in the water and churned it into foam, but presently the moon rose, and in its brilliant light I could see the females and the baby gathered at one end of the pool in a tight bunch, their heads gleaming above the surface of the water, their ears flicking to and fro. Now and again they would open their mouths and bray, rather in the manner of a Greek chorus. They were watching with interest both the old male and the young who were in the shallows at the centre of the pool. The water reached up only to their tummies, and their great barrel-shaped bodies and the rolls of fat under their chins gleamed as though they had been oiled. They were facing each other with lowered heads, snorting like a couple of steam-engines. Suddenly the young male lifted his great head, opened his mouth so that his teeth flashed in the moonlight, gave a prolonged and blood-curdling bray, and, just as he was finishing, the old male rushed at him with open mouth and the incredible speed for such a bulky animal. The young male, equally quick, twisted to one side. The old male splashed in a welter of foam like some misshapen battle-ship, and was now going so fast that he could not stop. As he passed, the young male, with a terrible sideways chop of his huge jaws, bit him in the shoulder. The old male swerved round and charged again, and just as he reached his opponent the moon went behind a cloud. When it came out again, they were standing as I had first seen them, facing each other with lowered heads, snorting.

I sat on that sandbank for two hours, watching these great roly-poly creatures churning up the water and sand as they duelled in the shallows. As far as I could see, the old male was getting the worst of it, and I felt sorry for him. Like some once-great pugilist who had now grown flabby and stiff, he seemed to be fighting a battle which he knew was already lost. The young male, lighter and more agile, seemed to dodge him every time, and his teeth always managed to find their mark in the shoulder or neck of the old male. In the background the females watched with semaphoring ears, occasionally breaking into a loud lugubrious chorus which may have been sorrow for the plight of the old male, or delight at the success of the young one, but was probably merely the excitement of watching the fight. Eventually, since the fight did not seem as if it would end for several more hours, I paddled home to the village and went to bed.

I awoke just as the horizon was paling into dawn, and the hippos were quiet. Apparently the fight was over. I hoped that the old male had won, but I very much doubted it. The answer was given to me later that morning by one of my hunters; the corpse of the old male, he said, was about two miles downstream, lying where the current of the river had carried it into the curving arms of a sandbank. I went down to examine it and was horrified at the havoc the young male's teeth had wrought on the massive body. The shoulders, the neck, the great dewlaps that hung under the chin, the flanks and the belly: all were ripped and tattered, and the shallows around the carcase were still tinged with blood. The entire village had accompanied me, for such an enormous windfall of meat was a red-letter day for them. They stood silent and interested while I examined the old male's carcase, and when I had finished and walked away they poured over it like ants, screaming and pushing with excitement, vigorously wielding their knives and machetes. It seemed to me, watching the huge hippo's carcase disintegrate under the pile of hungry humans, that it was a heavy price to pay for love.

A notably romantic member of the human race is described as hot-blooded; yet in the animal world it is among the cold-blooded creatures that you find some of the best courtship displays. The average crocodile looks as though he would prove a pretty cold-blooded lover as he lies on the bank, watching with his perpetual, sardonic grin and unwinking eyes the passing pageant of river life. Yet when the time and the place and the lady are right, he will fling himself into battle for her hand; and the two males, snapping and thrashing, will roll over and over in the water. At last the winner, flushed with victory, proceeds to do a strange dance on the surface of the water, whirling round and round with his head and tail thrust into the air, bellowing like a foghorn in what is apparently the reptilian equivalent of an old-fashioned waltz.

It is among the terrapins or water-tortoises that we find an example of the 'treat 'em rough and they'll love you' school of thought. In one of these little reptiles the claws on the front flippers are greatly elongated. Swimming along, the male sees a suitable female and starts to head her off. He then beats her over the head with his long fingernails, an action so quickly performed that his claws are a mere blur. This does not seem to make the female suffer in any way; it may even give her pleasure. But at any rate, even a female terrapin cannot succumb at the very first sign of interest on the part of the male. She must play hard to get, even if only for a short time, and she therefore breaks away and continues swimming in the stream. The male, now roused to a frenzy, swims after her, heads her off again, backs her up to the bank and proceeds to give her another beating. And this may happen several times before the female agrees to take up housekeeping with him. Whatever one may say against this reptile, he is certainly no hypocrite; he starts as he means to go on. And the female does not appear to mind these somewhat hectic advances. In fact, she seems to find them a pleasant and rather original form of approach. But there is no accounting for tastes – even among human beings.

However, for bewildering variety and ingenuity in the management of their love affairs, I think one must give pride of place to the insects.

Taking the praying mantis – mind you, one look at their

faces and nothing would surprise you about their private lives. The small head, the large, bulbous eyes dominating a tiny, pointed face that ends in a little quivering moustache; and the eyes themselves, a pale watery straw colour with black cat-like pupils that give them a wild and maniacal look. Under the chest a pair of powerful, savagely barbed arms are bent in a permanent and hypocritical attitude of prayer, being ready at a moment's notice to leap out and crush the victim in an embrace as though he had been caught in a pair of serrated scissors. Another unpleasant habit of the mantis is the way it looks at things, for it can turn its head to and fro in the most human manner and, if puzzled, will cock its little chinless face on one side, staring at you with wild eyes. Or, if you walk behind it, it will peer at you over its shoulder with an un-pleasant air of expectancy. Only a male mantis, I feel, could see anything remotely attractive in the female, and you would think he would be sensible enough not to trust a bride with a face like that. But no, I have seen one, his heart overflowing with love, clasp a female passionately, and while they were actually consummating their marriage his spouse leaned ten-derly over her shoulder and proceeded to eat him, browsing with the air of a gourmet over his corpse still clasped to her back, while her whiskers quivered and twitched as each delicate, glistening morsel was savoured to the full.

Female spiders, of course, have this same rather anti-social habit of eating their husbands; and the male's approach to the web of the female is thus fraught with danger. If she happens to be hungry, he will hardly have a chance to get the first words of his proposal out, as it were, before he finds himself a neatly trussed bundle being sucked of his vital juices by the lady. In one such species of the spider, the male has worked out a method to make certain he can get close enough to the female to tickle and massage her into a receptive frame of mind, without being eaten. He brings her a little gift – it may be a bluebottle or something of the sort – neatly wrapped up in silk. While she is busily devouring this, he creeps up behind her and strokes her into a sort of trance with his legs. Sometimes, when the nuptials are over, he manages to get away, but in most cases he is eaten at the end of the honeymoon, for it appears that the only true way to a female spider's heart is through her stomach.

In another species of spider the male has evolved an even more brilliant device for subduing his tigerish wife. Having approached her, he then starts to massage her gently with his legs until, as is usual with female spiders, she enters a sort of hypnotic state. Then the male, as swiftly as he can, proceeds to bind her to the ground with a length of silken cord, so that, when she awakes from her trance in the marriage bed, she finds herself unable to turn her husband into a wedding breakfast until she has set about the tedious business of untying herself. This generally saves the husband's life.

But if you want a really exotic romance you need not go to the tropical jungle to find it: just go into your own back-garden and creep up on the common snail. Here you have a situation as complex as the plot of any modern novel, for snails are hermaphrodite, and so each one can enjoy the pleasure of both the male and female side of courtship and mating. But apart from this dual sex, the snail possesses something even more extraordinary, a small sack-like container in its body in which is manufactured a tiny leaf-shaped splinter of carbonate of lime, known as the love-dart. Thus when one snail – who, as I say, is both male and female – crawls alongside another snail, also male and female, the two of them indulge in the most curious courtship action. They proceed to stab each other with their love-darts, which penetrate deeply and are quite quickly dissolved in the body. It seems that this curious duel is not as painful as it appears; in fact, the dart sinking into its side seems to give the snail a pleasant feeling, perhaps an exotic tickling sensation. But, whatever it is it puts both snails into an enthusiastic frame of mind for the stern business of mating. I am no gardener, but if I were I would probably have a soft spot for any snails in my garden, even if they did eat my plants. Any creature who has dispensed with Cupid, who carries his own quiverful of arrows around with him is, in my opinion, worth any number of dull and sexless cabbages. It is an honour to have him in the garden.

ANIMAL ARCHITECTS

SOME time ago I received a small parcel from a friend of mine in India. Inside the box I found a note which read: 'I bet you don't know what this is.' Greatly intrigued, I lifted off the top layer of wrapping paper, and underneath I found what appeared to be two large leaves which had been rather inexpertly sewn together.

My friend would have lost his bet. As soon as I saw the large and rather amateur stitches, I knew what it was: the nest of a tailor bird, a thing I had always wanted to see. The two leaves were about six inches long, shaped rather like laurel leaves, and only the edges had been sewn together, so that it formed a sort of pointed bag. Inside the bag was a neat nest of grass and moss, and inside that were two small eggs. The tailor bird is quite small, about the size of a tit, but with a rather long beak. This is its needle. Having found the two leaves it likes, hanging close together, it then proceeds to sew them together, using fine cotton as thread. The curious thing about it is not so much that the tailor bird stitches the leaves together as that nobody seems to know where he finds the cotton material with which to do the sewing. Some experts insist that he weaves it himself, others that he has some source of supply that has never been discovered. As I say, the stitches were rather large and inartistic, but then how many people could make a success of sewing up two leaves, using only a beak as a needle?

Architecture in the animal world differs a great deal. Some animals, of course, have only the haziest idea of constructing a suitable dwelling, while others produce most complicated and delightful homes. It is strange that even among closely

related animals there should be such a wide variety of taste in the style, situation and size of the home and the choice of materials used in its construction.

In the bird world, of course, one finds homes of every shape and size. They range from the tailor-bird's cradle of leaves to the emperor penguin, who, with nothing but snow for his building, has dispensed with the idea of a nest altogether. The egg is simply carried on the top of the large flat foot, and the skin and feathers of the stomach form a sort of pouch to cover it. Then you have the edible swift who makes a fragile, cup-shaped nest of saliva and bits of twigs and sticks it to the wall of a cave. Among the weaver-birds of Africa, too, the variety of nests is bewildering. One species lives in a community which builds a nest half the size of a haystack, rather like a block of flats, in which each bird has its own nesting-hole. In these gigantic nests you sometimes get an odd variety of creatures living as well as the rightful occupants. Snakes are very fond of them; so are bush-babies and squirrels. One of these nests, if taken to pieces, might display an extraordinary assortment of inmates. No wonder that trees have been known to collapse under the weight of these colossal nests. The common weaver-bird of West Africa builds a neat, round nest, like a small basket woven from palm fibres. They also live in communities and hang their nests on every available branch of a tree, until it seems festooned with some extraordinary form of fruit. In the most human way the brilliant and shrill-voiced owners go about the business of courting, hatching the eggs, feeding their young, and bickering with their neighbours, so that the whole thing rather resembles an odd sort of council estate.

To construct their nests, the weaver-birds have become adept not only at weaving but at tying knots, for the nest is strapped very firmly to the branch and requires considerable force to remove it. I once watched a weaver-bird starting its nest, a fascinating performance. He had decided that the nest should hang from the end of a delicate twig half-way up a tree, and he arrived on the spot carrying a long strand of palm fibre in his beak. He alighted on the branch, which at once swung to and fro so that he had to flap his wings to keep his balance. When he was fairly steady he juggled with the palm fibre until he got to the centre of it. Then he tried to drape it over the branch, so that the two ends hung one side and the loop hung the other. The branch still swayed about, and twice he dropped the fibre and had to fly down to retrieve it, but at last he got it slung over the branch to his satisfaction. He then placed one foot on it to keep it in position and leaning forward precariously he pulled the two dangling ends from one side of the branch through the loop on the other and tugged it tight. After this he flew for some more fibre and repeated the performance. He went on in this way for the whole day, until by evening he had twenty or thirty pieces of fibre lashed to the branch, the ends dangling down like a beard.

Unfortunately I missed the following stages in the construction of this nest, and I next saw it empty, for the bird had presumably reared its young and moved off. The nest was flask-shaped – a small round entrance, guarded by a small porch of plaited fibre. I tried to pull the nest off the branch, but it was impossible, and in the end I had to break the whole branch off. Then I tried to tear the nest in half so that I could examine the inside. But so intricately interlaced and knotted were the palm fibres that it took me a long time and all my strength before I could do so. It was really an incredible construction, when you consider the bird had only its beak and its feet for tools.

When I went to Argentina four years ago I noticed that nearly every tree-stump or rail-post in the pampa was decorated with a strange earthenware construction about the size and shape of a football. At first, I believed they were termite nests, for they were very similar to a common feature of the landscape in West Africa. It was not until I saw, perched on top of one of them, a small tubby bird about the size of a robin with a rusty-red back and grey shirt-front that I realized they were the nests of the oven-bird.

As soon as I found an unoccupied nest, I carefully cut it in half and was amazed at the skill with which it had been built. Wet mud had been mixed with tiny fragments of dried grass, roots and hair to act as reinforcement. The sides of the nest were approximately an inch and a half thick. The outside had been left rough – unrendered, as it were – but the inside had been smoothed to a glass-like finish. The entrance to the nest was a small arched hole, rather like a church door, which led into a narrow passage-way that curved round the outer edge of the nest and eventually led into the circular nesting-chamber lined with a pad of soft roots and feathers. The whole thing rather resembled a snail shell.

Although I searched a large area, I was never lucky enough to find a nest that had been newly started, for it was fairly well into the breeding season. But I did find one half-completed. Oven-birds are very common in Argentina, and in the way they move and cock their heads on one side and regard you with their shining dark eyes, they reminded me very much of the English robin. The pair building this nest took no notice of me whatever, provided I remained at a distance of about twelve feet, though occasionally they would fly over to take a closer look at me, and after inspecting me with their heads on one side, they would flap their wings as though shrugging, and return to their building work. The nest, as I say, was half-finished: the base was firmly cemented on to a fence-post and the outer walls and inner wall of the passage-way were already some four or five inches high. All that remained now was for the whole thing to be covered with the domed roof.

The nearest place for wet mud was about half a mile away at the edge of a shallow lagoon. They would hop round the edge of the water in a fussy, rather pompous manner, testing the mud every few feet. It had to be of exactly the right consistency. Having found a suitable patch, they would hop about excitedly, picking up tiny rootlets and bits of grass until their beaks were full and they looked as though they had suddenly sprouted large walrus moustaches. They would carry these beakfuls of reinforcement down to the mud patch, and then, by skilful juggling, without dropping the material, pick up a large amount of mud as well. By a curious movement of the beak they matted the two materials together until their walrus moustaches looked distinctly bedraggled and mudstained. Then, with a muffled squeak of triumph, they flew off to the nest. Here the bundle was placed in the right position and pecked and trampled on and pushed until it had firmly adhered to the original wall. Then they entered the nest and smoothed off the new patch, using their beaks, their breasts and even the sides of their wings to get the required shining finish.

When only a small patch on the very top of the roof needed to be finished, I took some bright scarlet threads of wool down to the edge of the lagoon and scattered them around the place where the oven-birds gathered their material. On my next trip down there, to my delight, they had picked them up, and the result, a small russet bird apparently wearing a bright scarlet moustache, was quite startling. They incorporated the wool into the last piece of building on the nest, and it was, I feel sure, the only Argentinian oven-bird's nest on the pampa flying what appeared to be a small red flag at half-mast.

If the oven-bird is a master-builder, whose nest is so solid that it takes several blows of a hammer to demolish it, members of the pigeon family go to the opposite extreme. They have absolutely no idea of proper nest-making. Four or five twigs laid across a branch: that is the average pigeon's idea of a highly complicated structure. On this frail platform the eggs, generally two, are laid. Every time the tree sways in the wind this silly nest trembles and shakes and the eggs almost fall out. How any pigeon ever reaches maturity is a mystery to me.

I knew that pigeons were stupid and inefficient builders, but I never thought that their nests might prove an irritating menace to a naturalist. When I was in Argentina I learned differently. On the banks of a river outside Buenos Aires I found a small wood. The trees, only about thirty feet high, were occupied by what might almost be called a pigeon colony. Every tree had about thirty or forty nests in it. Walking underneath the branches you could see the fat bellies of the young, or the gleam of the eggs, through the carelessly arranged twigs. The nests looked so insecure that I felt like walking on tip-toe for fear that my footsteps would destroy the delicate balance.

In the centre of the wood I found a tree full of pigeons' nests but for some odd reason devoid of pigeons. At the very top of the tree I noticed a great bundle of twigs and leaves which was obviously a nest of some sort and equally obviously not a pigeon's nest. I wondered if it was the occupant of this rather untidy bundle of stuff that had made the pigeons desert all the nests in the tree. I decided to climb and see if the owner was at home. Unfortunately, it was only when I had started to climb that I realized my mistake, for nearly every pigeon's nest in the tree contained eggs, and as I made my way slowly up the branches my movements created a sort of waterfall of pigeon eggs which bounced and broke against me, smearing my coat and trousers with yolk and bits of shell. I would not have minded this so much, but every single egg was well and truly addled, and by the time I had reached the top of the tree, hot and sweating, I smelt like a cross between a tannery and a sewage farm. To add insult to injury, I found that the occupant of the nest I had climbed up to was out, so I had gained nothing by my climb except a thick coating of egg and a scent that would have made a skunk envious. Laboriously I climbed down the tree again, looking forward to the moment when I would reach ground and could light a cigarette, to take the strong smell of rotten egg out of my nostrils. The ground under the tree was littered with broken eggs tastefully interspersed with the bodies of a few baby pigeons in a decomposed condition. I made my way out into the open as quickly as possible. With a sigh of relief, I sat down and reached into my pocket for my cigarettes. I drew them out dripping with egg-yolk. At some point during my climb, by some curious chance, an egg had fallen into my pocket and broken. My cigarettes were ruined. I had to walk two miles home without a smoke, breathing in a strong aroma of egg and looking as though I had rather unsuccessfully taken part in an omelette-making competition. I have never really liked pigeons since then.

Mammals, on the whole, are not such good builders as the birds, though, of course, a few of them are experts. The badger, for example, builds the most complicated burrow, which is sometimes added to by successive generations until the whole thing resembles an intricate underground system with passages, culs-de-sac, bedrooms, nurseries and feeding-quarters. The beaver, too, is another master-builder, constructing his lodge half in and half out of the water: thick walls of mud and logs with an underground entrance, so that he can get in and out even when the surface of the lake is iced over. Beavers also build canals, so that when they have to fell a tree some distance inland for food or repair work on their dam, they can float it down the canal to the main body of water. Their dams are, of course, masterpieces – massive constructions of mud and logs, welded together, stretching sometimes many hundreds of yards. The slightest breach in these is frantically repaired by the beavers, for fear that the water might drain away and leave their lodge with its door no longer covered by water, an easy prey to any passing enemy. What with their home, their canals, and their dams, one has the impression that the beaver must be a remarkably intelligent and astute animal. Unfortunately, however, this is not the case. It appears that the desire to build a dam is an urge which no self-respecting beaver can repress even when there is no need for the construction, and when kept in a large cement pool they will solemnly and methodically run a dam across it to keep the water in.

But, of course, the real master-architects of the animal world are, without a doubt, the insects. You need only look at the beautiful mathematical precision with which a common-or-garden honeycomb is built. Insects seem capable of building the most astonishing homes from a vast array of materials – wood, paper, wax, mud, silk, and sand – and they differ just as widely in their design. In Greece, when I was a boy, I used to spend hours searching mossy banks for the nest of the trap-door-spiders. These are one of the most beautiful and astonishing pieces of animal architecture in the world. The spider itself, with its legs spread out, would just about cover a two-shilling piece and looks as though it has been made out of highly polished chocolate. It has a squat fat body and rather short legs, and does not look at all the sort of creature you would associate with delicate construction work. Yet these rather clumsy-looking spiders sink a shaft into the earth of a bank about six inches deep and about the diameter of a shilling. This is carefully lined, so that when finished it is like a tube of silk. Then comes the most important part, the trapdoor. This is circular and with a neatly bevelled edge, so that it fits securely into the mouth of the tunnel. It is then fixed with a silken hinge, and the outside of it camouflaged with springs of moss or lichen; it is almost indistinguishable from the surrounding earth when closed. If the owner is not at home and you flip back the door, you will see on its silken underside a series of neat little black pinpricks. These are the handles, so to speak, in which the spider latches her claws to hold the door firmly shut against intruders. The only person, I think, who would not be amazed at the beauty of a trapdoor-spider's nest is the male trapdoor-spider himself, for once he has lifted the trapdoor and entered the silken shaft, it is for him both a tunnel of love and death. Once having gone down into the dark interior and mated with the female, he is promptly killed and eaten by her.

One of my first experiences with animal architects was when I was about ten years old. At that time I was extremely interested in freshwater biology and used to spend most of my spare time dredging about in ponds and streams, catching the minute fauna that lived there and keeping them in large jam-jars in my bedroom. Among other things, I had one jam-jar full of caddis larvae. These curious caterpillar-like creatures encase themselves in a sort of silken cocoon with one end open, and then decorate the outside of the cocoon with whatever materials they think will produce the best camouflage. The caddis I had were rather dull, for I had caught them in a very stagnant pool. They had merely decorated the outside of their cocoons with little bits of dead water-plant.

I had been told, however, that if you remove a caddis larva from its cocoon and place it in a jar of clean water, it would spin itself a new cocoon and decorate the outside with whatever materials you cared to supply. I was a bit sceptical about this, but decided to experiment. I took four of my caddis larvae and very carefully removed them, wriggling indignantly from their cocoons. Then I placed them in a jar of clean water and lined the bottom of the jar with a handful of tiny bleached seashells. To my astonishment and delight the creatures did exactly what my friend said they would do, and by the time the larvae had finished the new cocoons were like a filigree of seashells.

I was so enthusiastic about this that I gave the poor creatures a rather hectic time of it. Every now and then I would force them to manufacture new cocoons decorated with more and more improbable substances. The climax came with my discovery that by moving the larvae to a new jar with a new substance at the bottom when they were half-way through building operations, you could get them to build a particoloured cocoon. Some of the results I got were very odd. There was one, for example, who had half his cocoon magnificently arrayed in seashells and the other half in bits of charcoal. My greatest triumph, however, lay in forcing three of them to decorate their cocoons with fragments of blue glass, red brick and white seashells. Moreover, the materials were put on in stripes – rather uneven stripes, I grant you, but stripes nevertheless.

Since then I have had a lot of animals of which I have been proud, but I never remember feeling quite the same sort of satisfaction as I did when I used to show off my red, white and blue caddis larvae to my friends. I think the poor creatures were really rather relieved when they could hatch out and fly away and forget about the problems of cocoon-building.

ANIMAL WARFARE

I REMEMBER once lying on a sun-drenched hillside in Greece – a hillside covered with twisted olive-trees and myrtle bushes – and watching a protracted and bloody war being waged within inches of my feet. I was extremely lucky to be, as it were, war correspondent for this battle. It was the only one of its kind I have ever seen and I would not have missed it for the world.

The two armies involved were ants. The attacking force was a shining, fierce red, while the defending army was as black as coal. I might quite easily have missed this if one day I had not noticed what struck me as an extremely peculiar ants' nest. It contained two species of ants, one red and one black, living together on the most amicable terms. Never having seen two species of ants living in the same nest before, I took the trouble to check up on them, and discovered that the red ones, who were the true owners of the nest, were known by the resounding title of the blood-red slave-makers, and the black ones were in fact their slaves who had been captured and placed in their service while they were still eggs. After reading about the habits of the slave-makers, I kept a cautious eye on the nest in the hope of seeing them indulge in one of their slave raids. Several months passed and I began to think that either these slave-makers were too lazy or else they had enough slaves to keep them happy.

The slave-makers' fortress lay near the roots of an olive-tree, and some thirty feet farther down the hillside was a nest of black ants. Passing this nest one morning, I noticed several of the slave-makers wandering about within a yard or so of it, and I stopped to watch. There were perhaps thirty or forty of them, spread over quite a large area. They did not appear to be foraging for food, as they were not moving with their normal brisk inquisitiveness. They kept wandering round in vague circles, occasionally climbing a grass blade and standing pensively on its tip, waving their antennae. Periodically, two of them would meet and stand there in what appeared to be animated conversation, their antennae twitching together. It was not until I had watched them for some time that I realized what they were doing. Their wanderings were not as aimless as they appeared, for they were quartering the ground very thoroughly like a pack of hunting-dogs, investigating every bit of the terrain over which their army would have to travel. The black ants seemed distinctly ill at ease. Occasionally one of them would meet one of the slave-makers and would turn tail and run back to the nest to join one of the many groups of his relatives who were gathered in little knots, apparently holding a council of war. This careful investigation of the ground by the scouts of the slave-makers' army continued for two days, and I had begun to think that they had decided the black ants' city was too difficult to attack. Then I arrived one morning to find that the war had started.

The scouts, accompanied by four or five small platoons, had now moved in closer to the black ants, and already several skirmishes were taking place within two or three feet of the nest. Black ants were hurling themselves on the red ones with almost hysterical fervour, while the red ones were advancing slowly but inexorably, now and then catching a black ant and with a swift, savage bite piercing it through the head or the thorax with their huge jaws.

Half-way up the hillside the main body of the slave-makers' army was marching down. In about an hour they had got within four or five feet of the black ants' city, and here, with a beautiful military precision which was quite amazing to watch, they split into three columns. While one column marched directly on the nest the other two spread out and proceeded to execute a flanking or pincer movement. It was fascinating to watch. I felt I was suspended in some miraculous way above the field of battle of some old military campaign – the battle of Waterloo or some similar historic battle. I could see at a glance the disposition of the attackers and the defenders; I could see the columns of reinforcements hurrying up through the jungle of grass; see the two outflanking columns of slave-makers moving nearer and nearer to the nest, while the black ants, unaware of their presence, were concentrating on fighting off the central column. It was quite obvious to me that unless the black ants very soon realized that they were being encircled, they had lost all hope of survival. I was torn between a desire to help the black ants in some way and a longing to leave things as they were and see how matters developed. I did pick up one of the black ants and place him near the encircling red-ant column, but he was set upon and killed rapidly, and I felt quite guilty.

Eventually, however, the black ants suddenly became aware of the fact that they were being neatly surrounded. Immediately they seemed to panic; numbers of them ran to and fro aimlessly, some of them in their fright running straight into the red invaders and being instantly killed. Others, however, seemed to keep their heads, and they rushed down into the depths of the fortress and started on the work of evacuating the eggs, which they brought up and stacked on the side of nest farthest away from the invaders. Other members of the community then seized the eggs and started to rush them away to safety. But they had left it too late.

The encircling columns of slave-makers, so orderly and neat, now suddenly burst their ranks and spread over the whole area, like a scuttling red tide. Everywhere there were knots of

struggling ants. Black ones, clasping eggs in their jaws, were pursued by the slave-makers, cornered and then forced to give up the eggs. If they showed fight, they were immediately killed; the more cowardly, however, saved their lives by dropping the eggs they were carrying as soon as a slave-maker hove in sight. The whole area on and around the nest was littered with dead and dying ants of both species, while between the corpses the black ants ran futilely hither and thither, and the slave-makers gathered the eggs and started on the journey back to their fortress on the hill. At that point, very reluctantly, I had to leave the scene of battle, for it was getting too dark to see properly.

Early next morning I arrived at the scene again, to find the war was over. The black ants' city was deserted, except for the dead and injured ants littered all over it. Neither the black nor the red army were anywhere to be seen. I hurried up to the red ants' nest and was just in time to see the last of the army arrive there, carrying their spoils of war, the eggs, carefully in their jaws. At the entrance to the nest their black slaves greeted them excitedly, touching the eggs with their antennae and scuttling eagerly around their masters, obviously full of enthusiasm for the successful raid on their own relations that the slave-makers had achieved. There was something unpleasantly human about the whole thing.

It is perhaps unfair to describe animals as indulging in warfare, because for the most part they are far too sensible to engage in warfare as we know it. The exceptions are, of course, the ants, and the slave-makers in particular. But for most other creatures warfare consists of either defending themselves against an enemy, or attacking something for food.

After watching the slave-makers wage war I had the greatest admiration for their military strategy, but it did not make me like them very much. In fact, I was delighted to find that there existed what might be described as an underground movement bent on their destruction; the ant-lions. An adult ant-lion is very like a dragonfly, and looks fairly innocent. But in its childhood, as it were, it is a voracious monster that has evolved an extremely cunning way of trapping its prey, most of which consists of ants.

The larva is round-bodied, with a large head armed with great pincer-like jaws. Picking a spot where the soil is loose and sandy, it buries itself in the earth and makes a circular depression like the cone of a volcano. At the bottom of this, concealed by sand, the larva waits for its prey. Sooner or later an ant comes hurrying along in that preoccupied way so typical of ants, and blunders over the edge of the ant-lion's cone. It immediately realizes its mistake and tries to climb out again, but it finds this difficult, for the sand is soft and gives way under its weight. As it struggles futilely at the rim of this volcano it dislodges grains of sand which trickle down inside the cone and awake the deadly occupant that lurks there. Immediately the ant-lion springs into action. Using its great head and jaws like a steam-shovel, it shoots a rapid spray of sand grains at the ant, still struggling desperately to climb over the lip of the volcano. The earth sliding away from under its claws, knocked off its balance by this stream of sand and unable to regain it, the ant rolls down to the bottom of the cone where the sand parts like a curtain and it is enfolded lovingly in the great curved jaws of the ant-lion. Slowly, kicking and struggling, it disappears, as though it were being sucked down by quicksand, and within a few seconds the cone is empty, while below the innocent-looking sand the ant-lion is sucking the vital juices out of its victim.

Another creature that uses this sort of machine-gunning to bring down its prey is the archer-fish. This is a rather handsome creature found in the streams of Asia. It has evolved a most ingenious method of obtaining its prey, which consists of flies, butterflies, moths, and other insects. Swimming slowly along under the surface it waits until it sees an insect alight on a twig or leaf overhanging the water. Then the fish slows down and approaches cautiously. When it is within range it stops, takes aim, and then suddenly and startlingly spits a stream of tiny water droplets at its prey. These travel with deadly accuracy, and the startled insect is knocked off its perch and into the water below, and the next minute the fish swims up beneath it, there is a swirl of water and a gulp, and the insect has vanished for ever.

I once worked in a pet-shop in London, and one day, with a consignment of other creatures, we received an archer-fish. I was delighted with it, and with the permission of the manager I wrote out a notice describing the fish's curious habits, arranged the aquarium carefully, put the fish inside and placed it in the window as the main display. It proved very popular, except that people wanted to see the archer-fish actually taking his prey, and this was not easy to manage. Eventually I had a brainwave. A few doors down from us was a fish shop, and I saw no reason why we should not benefit from some of their surplus bluebottles. So I suspended a bit of very smelly meat over the archer-fish's aquarium and left the door of the shop open. I did this without the knowledge of the manager. I wanted it to be a surprise for him.

It was certainly a surprise.

By the time he arrived, there must have been several thousand bluebottles in the shop. The archer-fish was having the time of his life, watched by myself inside the shop and fifty or sixty people on the pavement outside. The manager arrived neck and neck with a very unzoological policeman, who wanted to know the meaning of the obstruction outside. To my surprise the manager, instead of being delighted with my ingenious window display, tended to side with the policeman. The climax came when the manager, leaning over the aquarium to unfasten the bit of meat that hung above it, was hit accurately in the face by a stream of water which the fish had just released in the hope of hitting a particularly succulent bluebottle. The manager never referred to the incident again, but the next day the archer-fish disappeared, and it was the last time I was allowed to dress the window.

Of course, one of the favourite tricks in animal warfare is for some harmless creature to persuade a potential enemy that it is really a hideous, ferocious beast, best left alone. One of the most amusing examples of this I have seen was given to me by a sun bittern when I was collecting live animals in British Guiana. This slender bird, with a delicate, pointed beak and slow, stately movements, had been hand-reared by an Indian and was therefore perfectly tame. I used to let it wander freely round my camp during the day and lock it in a cage only at night. Sun bitterns are clad in lovely feathering that has all the hints of an autumn woodland, and sometimes when this bird stood unmoving against a background of dry leaves she seemed to disappear completely. As I say, she was a frail, dainty little bird who, one would have thought, had no defence of any sort against an enemy. But this was not the case.

Three large and belligerent hunting-dogs followed their master into camp one afternoon, and before long one of them spotted the sun bittern, standing lost in meditation on the edge of the clearing. He approached her, his ears pricked, growling softly. The other two quickly joined him, and the three of them bore down on the bird with a swaggering air. The bird let them get within about four feet of her before deigning to notice them. Then she turned her head, gave them a withering stare and turned round to face them. The dogs paused, not quite sure what to do about a bird that did not run squawking at their approach. They moved closer. Suddenly the bittern ducked her head and spread her wings, so that the dogs were presented with a fan of feathers. In the centre of each wing was a beautiful marking, not noticeable when the wings were closed, which looked exactly like the two eyes of an enormous owl glaring at you. The whole transformation was done so slickly, from a slim meek little bird to something that resembled an infuriated eagle owl at bay, that the dogs were taken completely by surprise. They stopped their advance, took one look at the shivering wings and then turned tail and fled. The sun bittern shuffled her wings back into place, preened a few of her breast feathers that had become disarranged and fell to meditating again. It was obvious that dogs did not trouble her in the slightest.

Some of the most ingenious methods of defence in the animal world are displayed by insects. They are masters of the art of disguise, of setting traps, and other methods of defence and attack. But, certainly, one of the most extraordinary is the bombardier beetle.

I was once the proud owner of a genuine wild black rat which I had caught when he was a half-grown youngster. He was an extremely handsome beast with his shining ebony fur and gleaming black eyes. He divided his time equally between cleaning himself and eating. His great passion was for insects of any shape or size: butterflies, praying mantis, stick-insects, cockroaches, they all went the same way as soon as they were put into his cage. Not even the largest praying mantis stood a chance against him, though they would occasionally manage to dig their hooked arms into his nose and draw a bead of blood before he scrunched them up. But one day I found an insect which got the better of him. It was a large, blackish beetle which had been sitting reflecting under a stone that I had inquisitively turned over; and, thinking it would make a nice titbit for my rat, I put it in a matchbox in my pocket. When I arrived home I pulled the rat out of his sleeping-box, opened the matchbox and shook the large succulent beetle on to the floor of his cage. Now the rat had two methods of dealing with insects, which varied according to their kind. If they were as fast-moving and as belligerent as a mantis, he would rush in and bite as quickly as possible in order to destroy it, but with anything harmless and slow, like a beetle, he would pick it up in his paws and sit scrunching it up as though it were a piece of toast.

Seeing this great fat delicacy wandering rather aimlessly around on the floor of his cage, he trotted forward, rapidly seized it with his little pink paws and then sat back on his haunches with the air of a gourmet about to sample the first truffle of the season. His whiskers twitched in anticipation as he lifted the beetle to his mouth, and then a curious thing happened. He uttered the most prodigious sniff, dropped the beetle and leaped backwards as though he had been stung, and sat rubbing his paws hastily over his nose and face. At first I thought he had merely been taken with a sneezing fit just as he was about to eat the beetle. Having wiped his face, he again approached it, slightly more cautiously this time, picked it up and lifted it to his mouth. Then he uttered a strangled snort, dropped it as though it were red-hot and sat wiping his face indignantly. The second experience had obviously been enough for him, for he refused to go near the beetle after that; in fact he seemed positively scared of it. Every time it ambled round to the corner of the cage where he was sitting, he would back away hurriedly. I put the beetle back in the matchbox and took it inside to identify it and it was only then that I discovered that I had offered my unfortunate rat a bombardier beetle. Apparently the beetle, when attacked, squirts out a liquid which, on reaching the air, explodes with a tiny crack and forms a sort of pungent and unpleasant gas, sufficiently horrible to make any creature who has experienced it leave the bombardier beetle severely alone in future.

I felt rather sorry for my black rat. It was, I felt, an unfortunate experience to pick up what amounted to a particularly delicious dinner, only to have it suddenly turn into a gas attack in your paws. It gave him a complex about beetles, too, because for days afterwards he would dash into his sleeping-box at the sight of one, even a fat and harmless dung-beetle. However, he was a young rat, and I suppose he had to learn at some time or another that one cannot judge by appearances in this life.

ANIMAL INVENTORS

I ONCE travelled back from Africa on a ship with an Irish captain who did not like animals. This was unfortunate, because most of my luggage consisted of about two hundred-odd cages of assorted wild life, which were stacked on the forward well deck. The captain (more out of devilment than anything else, I think) never missed a chance of trying to provoke me into an argument by disparaging animals in general and my animals in particular. But fortunately I managed to avoid getting myself involved. To begin with, one should never argue with the captain of a ship, and to argue with a captain who was also an Irishman was simply asking for trouble. However, when the voyage was drawing to an end, I felt the captain needed a lesson and I was determined to teach him one if I could.

One evening when we were nearing the English Channel, the wind and rain had driven us all into the smoking-room, where we sat and listened to someone on the radio giving a talk on radar, which in those days was still sufficiently new to be of interest to the general public. The captain listened to the talk with a gleam in his eye, and when it had finished he turned to me.

'So much for your animals,' he said, 'they couldn't produce anything like that, in spite of the fact that, according to you, they're supposed to be so clever.'

By this simple statement the captain had played right into my hands, and I prepared to make him suffer.

'What will you bet,' I inquired, 'that I can't describe at least two great scientific inventions and prove to you that the principle was being used in the animal world long before man ever thought of it?'

'Make it four inventions instead of two and I'll bet you a bottle of whisky,' said the captain, obviously feeling he was on to a good thing. I agreed to this.

'Well,' said the captain smugly, 'off you go.'

'You'll have to give me a minute to think,' I protested.

'Ha,' said the captain triumphantly, 'you're stuck already.'

'Oh, no,' I explained, 'it's just that there are so many examples I'm not sure which to choose.'

The captain gave me a dirty look.

'Why not try radar, then?' he inquired sarcastically.

'Well, I could,' I said, 'but I really felt it was too easy. However, since you choose it, I suppose I'd better.'

It was fortunate for me that the captain was no naturalist; otherwise he would never have suggested radar. It was a gift, from my point of view, because I simply described the humble bat.

Many people must have been visited by a bat in their drawing-room or bedroom at one time or another, and if they have not been too scared of it, they will have been fascinated by its swift, skilful flight and the rapid twists and turns with which it avoids all obstacles, including objects like shoes and towels that are sometimes hurled at it. Now, despite the old saying, bats are not blind. They have perfectly good eyes, but these are so tiny that they are not easily detected in the thick fur. Their eyes, however, are certainly not good enough for them to perform some of the extraordinary flying stunts in which they indulge. It was an Italian naturalist called Spallanzani, in the eighteenth century, who first started to investigate the flight of bats, and by the unnecessarily cruel method of blinding several bats he found that they could still fly about unhampered, avoiding obstacles as though they were uninjured. But how they managed to do this he could not guess.

It was not until fairly recently that this problem was solved, at least partially. The discovery of radar, the sending out of sound-waves and judging the obstacles ahead by the returning echo, made some investigators wonder if this was not the system employed by bats. A series of experiments was conducted, and some fascinating things were discovered. First of all, some bats were blindfolded with tiny pieces of wax over their eyes, and as usual they had no difficulty in flying to and fro without hitting anything. Then it was found that if they were blindfolded and their ears were covered they were no longer able to avoid collisions, and, in fact, did not seem at all keen on flying in the first place. If only one ear was covered they could fly with only moderate success, and would frequently hit objects. This showed that bats could get information about the obstacles ahead by means of sound-waves reflected from them. Then the investigators covered the noses and mouths of their bats, but left the ears uncovered, and again the bats were unable to fly without collision. This proved that the nose, ears, and mouth all played some part in the bat's radar system. Eventually, by the use of extremely delicate instruments, the facts were discovered. As the bat flies along, it emits a continuous succession of supersonic squeaks, far too high for the human ear to pick up. They give out, in fact, about thirty squeaks a second. The echoes from these squeaks, bouncing off the obstacles ahead, return to the bat's ears and, in some species, to the curious fleshy ridges round the creature's nose, and the bat can thus tell what lies ahead, and how far away it is. It is, in fact, in every detail the principle of radar. But one thing rather puzzled the investigators: when you are transmitting sound-waves on radar, you must shut off your receiver when you are actually sending out the sound, so that you receive only the echo. Otherwise the receiver would pick up both the sound transmitted and the echo back, and the result would be a confused jumble. This might be possible on electrical apparatus, but they could not imagine how the bats managed to do it. It was then discovered that there was a tiny muscle in the bat's ear that did the job. Just at the moment the bat squeaks, this muscle contracts and puts the ear out of action. The squeak over, the muscle relaxes and the ear is ready to receive the echo.

But the amazing thing about this is not that bats have this private radar system – for after a while very little surprises one in Nature – but that they should have had it so long before man did. Fossil bats have been found in early Eocene rocks, and they differed very little from their modern relatives. It is possible, therefore, that bats have been employing radar for something like fifty million years. Man has possessed the secret for about twenty.

It was quite obvious that my first example had made the captain think. He did not seem quite so sure of winning the bet. I said that my next choice would be electricity, and this apparently cheered him up a bit. He laughed in a disbelieving way, and said I would have a job to persuade him that animals had electric lights. I pointed out that I had said nothing about electric lights, but merely electricity, and there were several creatures that employed it. There is, for example, the electric-ray or torpedo-fish, a curious creature that looks rather like a frying-pan run over by a steam-roller. These fish are excessively well camouflaged: not only does their colouring imitate the sandy bottom but they have also the annoying habit of half-burying themselves in the sand, which renders them really invisible. I remember once seeing the effect of this fish's electric organs, which are large and situated on its back. I was in Greece at the time, and was watching a young peasant boy fishing in the shallow waters of a sandy bay. He was wading up to his knees in the clear waters, holding in his hand a three-pronged spear such as the fishermen used for night-fishing. As he made his way round the bay, he was having quite a successful time: he had speared several large fish and a young octopus which had been concealed in a small group of rocks. As he came opposite where I was sitting a curious and rather startling thing happened. One minute he was walking slowly forward, peering down intently into the water, his trident at the ready; the next minute he had straightened up as stiffly as a guardsman and projected himself out of the water like a rocket, uttering a yell that could have been heard half a mile away. He fell back into the water with a splash and immediately uttered another and louder scream and leapt up again. This time he fell back into the water and seemed unable to regain his feet, for he struggled out on to the sand, half crawling, half dragging himself. When I got down to where he lay, I found him white and shaking, panting as though he had just run half a mile. How much of this was due to shock and how much to the actual effect of the electricity I could not tell, but at any rate I never again went bathing in that particular bay.

Probably the most famous electricity-producing creature is the electric-eel which, strangely enough, is not an eel at all but a species of fish that looks like an eel. These long, black creatures live in the streams and rivers of South America and can grow to eight feet in length and the thickness of a man's thigh. No doubt a lot of stories about them are grossly exaggerated, but it is possible for a big one to shock a horse fording a river strongly enough to knock down the animal.

When I was collecting animals in British Guiana I very much wanted to catch some electric-eels to bring back to this country. At one place where we were camped the river was full of them, but they lived in deep caves hollowed out in the rocky shores. Most of these caves communicated with the air by means of round pot-holes that had been worn by the flood waters, and in the cave beneath each pot-hole lived an electric-eel. If you made your way to a pot-hole and stamped heavily with your shoes it would annoy the eel into replying with a strange purring grunt, as though a large pig were entombed beneath your feet.

Try as I would I did not manage to catch one of these eels. Then one day my partner and I, accompanied by two Indians, went for a trip to a village a few miles away, where the inhabitants were great fishermen. We found several animals and birds in the village which we purchased from them, including a tame tree-porcupine. Then, to my delight, someone appeared with an electric-eel in a rather insecure fish-basket. Having bargained for and bought these creatures, including the eel, we piled them into the canoe and set out for home. The porcupine sat in the bow, apparently very interested in the scenery, and in front of him lay the eel in its basket. We were half-way home when the eel escaped.

We were first made aware of this by the porcupine. He was, I think, under the impression that the eel was a snake, for he galloped down from the bows and endeavoured to climb on to my head. Struggling to evade the porcupine's prickly embrace, I suddenly saw the eel wriggling determinedly towards me, and indulged in a feat which I would not have believed possible. I leapt into the air from a sitting position, clasping the porcupine to my bosom, and landed again when the eel had passed, without upsetting the canoe. I had a very vivid mental picture of what had happened to the young peasant who had trodden on the torpedo-fish, and I had no intention of indulging in a similar experience with an electric-eel. Luckily none of us received a shock from the eel, for while we were trying to juggle it back into its basket it wriggled over the side of the canoe and fell into the river. I cannot say any of us were really sorry to see it go.

I remember once feeding an electric-eel that lived in a large tank in a zoo, and it was quite fascinating to watch his method of dealing with his prey. He was about five feet long and could cope adequately with a fish of about eight or ten inches in length. These had to be fed to him alive, and as their death was instantaneous, I had no qualms about this. The eel seemed to know when it was feeding-time and he would be patrolling his tank with the monotonous regularity of a sentry outside Buckingham Palace. As soon as a fish was dropped into his tank he would freeze instantly and apparently watch it as it swam closer and closer. When it was within range, which was about a foot or so, he would suddenly appear to quiver all over as if a dynamo had started within his long dark length. The fish would be, as it were, frozen in its tracks; it was dead before you realized that anything was happening, and then very slowly it would tilt over and start floating belly uppermost. The eel would move a little closer, open his mouth and suck violently, and, as though he were an elongated vacuum-cleaner, the fish would disappear into him.

Having dispensed quite successfully, I thought, with electricity, I now turned my attention to another field: medicine. Anaesthetics, I said, would be my next example, and the captain looked if anything even more sceptical than before.

The hunting-wasp is the Harley Street specialist of the insect world, and he performs an operation which would give a skilled surgeon pause. There are many different species of hunting-wasp, but most of them have similar habits. For the reception of her young the female has to build a nursery out of clay. This is neatly divided into long cells about the circumference of a cigarette and about half its length. In these the wasp intends to lay her eggs. However, she has another duty to perform before she can seal them up, for her eggs will hatch into grubs, and they will then require food until such time as they are ready to undergo the last stage of their meta-morphosis into the perfect wasp. The hunting-wasp could stock her nursery with dead food, but by the time the eggs had hatched this food would have gone bad, so she is forced to evolve another method. Her favourite prey is the spider. Flying like some fierce hawk, she descends upon her un-suspecting victim and proceeds to sting it deeply and skilfully. The effect of this sting is extraordinary, for the spider is completely paralyzed. The hunting-wasp then seizes it and carries it off to her nursery where it is carefully tucked away in one of the cells and an egg laid on it. If the spiders are small, there may be anything up to seven or eight in a cell. Having satisfied herself that the food-supply is adequate for her youngsters, the wasp then seals up the cells and flies off. Inside this grisly nursery the spiders lie in an unmoving row, in some cases for as much as seven weeks. To all intents and purposes the spiders are dead, even when you handle them, and not even under a magnifying glass can you detect the faintest sign of life. Thus they wait, so to speak, in cold storage until the eggs hatch out and the tiny grubs of the hunting-wasp start browsing on their paralyzed bodies.

I think even the captain was a little shaken by the idea of being completely paralyzed while something consumed you bit by bit, so I hastily switched to something a shade more pleasant. It was, in fact, the most delightful little creature, and a most ingenious one – the water-spider. Only recently in his history has man been able to live under water for any length of time, and one of his first steps in this direction was the diving-bell. Thousands of years before this the water-spider had evolved his own method of penetrating this new world beneath the surface of the water. To begin with, he can quite happily swim below the surface of the water, wearing his equivalent of the aqualung in the shape of an air bubble which he traps beneath his stomach and between his legs, so that he may breathe under water. This alone is extraordinary, but the water-spider goes even further: he builds his home beneath the surface of the water, a web shaped like an inverted cup, firmly anchored to the water-weeds. He then proceeds to make several journeys to the surface, bringing with him air bubbles which he pushes into this dome-shaped web until it is full of them, and in this he can live and breathe as easily as if he were on land. In the breeding-season he picks out the house of a likely looking female and builds himself a cottage next door, and then, presumably being of a romantic turn of mind, he builds a sort of secret passage linking his house with that of his lady-love. Then he breaks down her wall, so that the air bubbles in each house intermix, and here in this strange underwater dwelling he courts the female, mates with her, and lives with her until the eggs are laid and hatched, and until their children, each carrying their little globule of air from their parents' home, swim out to start life on their own.

Even the captain seemed amused and intrigued by my story of the water-spider, and he was bound to admit, albeit reluctantly, that I had won my bet.

I suppose it must have been about a year later I was talking to a lady who had travelled on the same ship with the same captain.

'Wasn't he a delightful man?' she asked me. I agreed politely.

'He must have enjoyed having you on board,' she went on, 'because he was so keen on animals, you know. One night he

kept us all spellbound for *at least* an hour, telling us about all these scientific discoveries – you know, things like radar – and how animals have been employing them for years and years before man discovered them. Really it was fascinating. I told him he ought to write it up into a talk and broadcast it on the B.B.C.'

VANISHING ANIMALS

SOME time ago I was watching what must be the strangest group of refugees in this country, strange because they did not come here for the usual reasons, driven by either religious or political persecution from their own country. They came here quite by chance, and in doing so they were saved from extermination. They are the last of their kind, for in their country of origin their relatives were long ago hunted down, killed and eaten. They were, in fact, a herd of Père David deer.

Their existence was first discovered by a French missionary, one Father David, during the course of his work in China in the early eighteen hundreds. In those days China was as little known, zoologically speaking, as the great forests of Africa, and so Father David, who was a keen naturalist, spent his spare time collecting specimens of the flora and fauna to send back to the museum in Paris. In 1865 his work took him to Peking, and while he was there he heard a rumour that there was a strange herd of deer in the Imperial Hunting Park, just south of the city. This park had been for centuries a sort of combined hunting- and pleasure-ground for the Emperors of China, a great tract of land completely surrounded by a high wall forty-five miles long. It was strictly guarded by Tartar soldiers, and no one was allowed to enter or approach it. The French missionary was intrigued by the stories he heard about these peculiar deer, and he was determined that, guards or no guards, he was going to look inside the walled park and try to see the animals for himself. One day he got his opportunity and was soon lying up on top of the wall, looking down into the forbidden park and watching the various game animals feeding among the trees below him. Among them was a large herd of deer, and Father David realized that he was looking at an animal he had never seen before, and one which was, very probably, new to science.

Father David soon found out that the deer were strictly protected, and for anyone caught harming or killing them the sentence was death. He knew that any official request he might put forward for a specimen would be politely refused by the Chinese authorities, so he had to use other, less legal methods to get what he wanted. He discovered that the Tartar guards occasionally improved their rather sparse rations by the addition of a little venison; they were well aware what the penalty for their poaching would be if they were caught, and so, in spite of the missionary's pleadings, they refused to sell him the skins and antlers of the deer they killed, or indeed anything that might be evidence of their crime. However, Father David did not give up hope, and after a considerable time he was successful. He met some guards who were either braver or perhaps poorer than the rest, and they obtained for him two deer skins, which he triumphantly shipped off to Paris. As he had expected, the deer turned out to be an entirely new species, and so it was named, in honour of its discoverer, the Père David deer – Father David's deer.

Naturally, when zoos in Europe heard about this new kind of deer they wanted specimens for exhibition, and after protracted negotiations the Chinese authorities rather reluctantly allowed a few of the animals to be sent to the Continent. Although no one realized it at the time, it was this action that was to save the animals. In 1895, thirty years after the Père David deer first became known to the world, there were great floods around Peking; the Hun-Ho river overflowed its banks and caused havoc in the countryside, destroying the crops and bringing the population to near starvation. The waters also undermined the great wall round the Imperial Hunting Park.

Parts of it collapsed, and through these gaps the herd of Père David deer escaped into the surrounding countryside, where they were quickly killed and eaten by the hungry peasants. So the deer perished in China, and the only ones left were the handful of live specimens in the various zoos in Europe.

Shortly before this disaster overtook the deer in China, a small herd of them had arrived in England. The present Duke of Bedford's father had, on his estate at Woburn in Bedfordshire, a wonderful collection of rare animals, and he had been most anxious to try to establish a herd of this new Chinese deer there. He bought as many specimens as he could from the Continental zoos, eighteen in all, and released them in his park. To the deer this must have seemed like home from home, for they settled down wonderfully, and soon started to breed. Today, the herd that started with eighteen now numbers over a hundred and fifty animals, the only herd of Père David deer in the world.

When I was working at Whipsnade Zoo four newly born Père David deer were sent over from Woburn for us to hand-rear. They were delightful little things, with long gangling limbs over which they had no control and strange slanted eyes that gave them a distinctly Oriental appearance. To begin with, of course, they did not know what a feeding-bottle was for, and we had to hold them firmly between our knees and force them to drink. But they very soon got the hang of it, and within a few days we had to open the stable door with extreme caution if we did not want to be knocked flying by an avalanche of deer, pushing and shoving in an effort to get at the bottle first.

They had to be fed once during the night, at midnight, and again at dawn, and so we worked out a system of night duties, one week on, one week off, between four keepers. I must say that I rather enjoyed the night duties. To pick one's way through the moonlit park towards the stable where the baby deer were kept, you had to pass several of the cages and paddocks, and the occupants were always on the move. The bears, looking twice as big in the half-light, would be snorting to each other as they shambled heavily through the riot of brambles in their cage, and they could be persuaded to leave their quest for snails and other delicacies if one had a bribe of sugar-lumps. They would come and squat upright in the moonlight, like a row of shaggy, heavy-breathing Buddhas, their great paws resting on their knees. They would throw back their heads and catch the flying lumps of sugar and eat them with much scrunching and smacking of lips. Then, seeing that you had no more in your pockets, they would sigh in a long-suffering manner and shamble off into the brambles again.

At one point the path led past the wolf wood, two acres or so of pines, dark and mysterious, with the moonlight silvering the trunks and laying dark shadows along the ground through which the wolf pack danced on swift, silent feet, like a strange black tide, swirling and twisting among the trunks. As a rule they made no sound, but occasionally you would hear them panting gently, or the sudden snap of jaws and a snarl when one wolf barged against another.

Then you would reach the stable and light the lantern. The baby deer would hear you and start moving restlessly in their straw beds, bleating tremulously. As you opened the door they rushed forward, wobbling on their unsteady legs, sucking frantically at your fingers, the edge of your coat, and butting you suddenly in the legs with their heads, so that you were almost knocked down. Then came the exquisite moment when the teat was pushed into their mouths and they sucked frantically at the warm milk, their eyes staring, bubbles gathering like a moustache at the corners of their mouths. There is always a certain pleasure to be gained from

bottle-feeding a baby animal, if only from its wholehearted enthusiasm and concentration on the job. But in the case of these deer there was something else as well. In the flickering light of the lantern, while the deer sucked and slobbered over the bottles, occasionally ducking their heads and butting at an imaginary udder with their heads, I was very conscious of the fact that they were the last of their kind.

At Whipsnade I had to look after another group of animals which belonged to a species now extinct in the wild state, and they were some of the most charming and comic animals I have ever had anything to do with. They were a small herd of white-tailed gnus.

The white-tailed gnu is a weird creature to look at: if you can imagine an animal with the body and legs of a finely built pony, a squat blunt face with very wide-spaced nostrils, a heavy mane of white hair on its thick neck, and a long white sweeping plume of a tail. The buffalo horns curve outward and upwards over the eyes, and the animal peers at you from under them with a perpetually indignant and suspicious expression. If the gnu behaved normally, this appearance would not be so noticeable, but the animal does not behave normally. Anything but, in fact. Its actions can only be described, very inadequately, as a cross between bebop and ballet, with a bit of yoga thrown in.

In the mornings, when I went to feed them, it always took me twice as long as it should have done because the gnus would start performing for me, and the sight was so ludicrous that I would lose all sense of time. They would prance and twist and buck, gallop, rear and pirouette, and while they did so they would throw their slim legs out at extraordinary and completely un-anatomical angles, and swish and curve their long tails as a circus ringmaster uses his whip. In the middle of the wild dance they would suddenly stop dead and glare at me, uttering loud, indignant belching snorts at my laughter. I watched them dancing their swift, wild dance across the paddock and they reminded me, in their antics and attitudes, of some strange heraldic creature from an ancient coat-of-arms, miraculously brought to life, prancing and posturing on a field of green turf.

It is difficult to imagine how anyone had the heart to kill these agile and amusing antelopes. However, the fact remains that the early settlers in South Africa found in the white-tailed gnu a valuable source of food, and so the great herds of high-spirited creatures were slaughtered unmercifully. The antelope contributed to its own downfall in an unusual way. They are incorrigibly curious creatures, and so when they saw the ox-drawn waggons of the early settlers moving across the veldt they simply had to go and investigate. They would dance and gallop round the waggons in circles, snorting and kicking their heels, and then suddenly stopping to stare. Naturally, with these habits of running away and then stopping to stare before they were out of range, they were used by enterprising 'sportsmen' for rifle practice. So they were killed, and their numbers decreased so rapidly that it is amazing that they did not become extinct. Today there are under a thousand of these charming animals left alive, and these are split up into small herds on various estates in South Africa. If they were to become extinct, South Africa would have lost one of the most amusing and talented of her native fauna, an antelope whose actions could enliven any landscape, however dull.

Unfortunately, the Père David deer and the white-tailed gnu are not the only creatures in the world that are nearly extinct. The list of creatures that have vanished altogether, and others that have almost vanished, is a long and melancholy one. As man has spread across the earth he has wrought the most terrible havoc among the wild life by shooting, trapping, cutting and burning the forest, and by the callous and stupid introduction of enemies where there were no enemies before.

Take the dodo, for example, the great ponderous waddling pigeon, the size of a goose, that inhabited the island of Mauritius. Secure in its island home, this bird had lost the power of flight since there were no enemies to fly from, and, since there were no enemies, it nested on the ground in complete safety. But, as well as losing the power of flight, it seems to have lost the power of recognizing an enemy when it saw one, for it was apparently an extremely tame and confiding creature. Then man discovered the dodos' paradise in about 1507, and with him came his evil familiars: dogs, cats, pigs, rats and goats. The dodo surveyed these new arrivals with an air of innocent interest. Then the slaughter began. The goats ate the undergrowth which provided the dodo with cover; dogs and cats hunted and harried the old birds; while pigs grunted their way round the island, eating the eggs and young and the rats followed behind to finish the feast. By 1681 the fat, ungainly and harmless pigeon was extinct – as dead as the dodo.

All over the world the wild fauna has been whittled down steadily and remorselessly, and many lovely and interesting animals have been so reduced in numbers that, without protection and help, they can never re-establish themselves. If they cannot find sanctuary where they can live and breed undisturbed, their numbers will dwindle until they join the dodo, the quagga, and the great auk on the long list of extinct creatures.

Of course, in the last decade or so much has been done for the protection of wild life: sanctuaries and reserves have been started, and the reintroduction of a species into areas where it had become extinct is taking place. In Canada, for instance, beavers are now reintroduced into certain areas by means of aeroplane. The animal is put in a special box attached to a parachute, and when the plane flies over the area it drops the cage and its beaver passenger out. The cage floats down on the end of the parachute, and when it hits the ground it opens automatically and the beaver then makes its way to the nearest stream or lake.

But although much is being done, there is still a very great deal to do. Unfortunately, the majority of useful work in animal preservation has been done mainly for animals which are of some economic importance to man, and there are many obscure species of no economic importance which, although they are protected on paper, as it were, are in actual fact being allowed to die out because nobody, except a few interested zoologists, considers them important enough to spend money on.

As mankind increases year by year, and as he spreads farther over the globe burning and destroying, it is some small comfort to know that there are certain private individuals and some institutions who consider that the work of trying to save and give sanctuary to these harried animals is of some importance. It is important work for many reasons, but perhaps the best of them is this: man, for all his genius, cannot create a species, nor can he recreate one he has destroyed. There would be a dreadful outcry if anyone suggested obliterating, say, the Tower of London, and quite rightly so; yet a unique and wonderful species of animal which has taken hundreds of thousands of years to develop to the stage we see today, can be snuffed out like a candle without more than a handful of people raising a finger or a voice in protest. So, until we consider animal life to be worthy of the consideration and reverence we bestow upon old books and pictures and historic monuments, there will always be the animal refugee living a precarious life on the edge of extermination, dependent for existence on the charity of a few human beings.

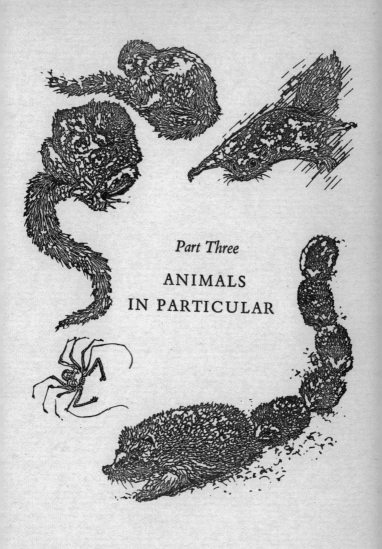

Part Three

ANIMALS
IN PARTICULAR

KEEPING wild animals as pets, whether on an expedition or in your own home, can be a tedious, irritating, and frustrating business, but it can also give you a great deal of pleasure. Many people have asked me why I like animals, and I have always found it a difficult question to answer. You might just as well ask me why I like eating. But, apart from the obvious interest and pleasure that animals give me, there is another aspect as well. I think that their chief charm lies in the fact that they have all the basic qualities of a human being but with none of the hypocrisy which is now apparently such an essential in the world of man. With an animal you do know more or less where you are: if it does not like you it tells you so in no uncertain manner; if it likes you, again it leaves you in no doubt. But an animal who likes you is sometimes a mixed blessing. Recently I had a pied crow from West Africa who, after six months' deliberation, during which time he ignored me, suddenly decided that I was the only person in the world for him. If I went near the cage he would crouch on the floor trembling in ecstasy, or bring me an offering (a bit of newspaper or a feather) and hold it out for me to take, all the while talking hoarsely to himself in a series of hiccuping cries and ejaculations. This was all right, but as soon as I let him out of his cage he would fly on to my head and perch there, first digging his claws firmly into my scalp, then decorating the back of my jacket with a nice moist dropping and finally proceeding to give me a series of love pecks on the head. As his beak was three inches long and extremely sharp, this was, to say the least, painful.

Of course, you have to know where to draw the line with animals. You can let pet-keeping develop into eccentricity if you are not careful. I drew the line last Christmas. For a present I decided to buy my wife a North American flying-squirrel, a creature which I had always wanted to possess myself, and which I was sure she would like. The animal duly arrived, and we were both captivated by it. As it seemed extremely nervous, we thought it would be a good idea to keep it in our bedroom for a week or two, so that we could talk to it at night when it came out, and let it grow used to us. This plan would have worked quite well but for one thing. The squirrel cunningly gnawed its way out of the cage and took up residence behind the wardrobe. At first this did not seem too bad. We could sit in bed at night and watch it doing acrobatics on the wardrobe, scuttling up and down the dressing-table, carrying off the nuts and apple we had left there for it. Then came New Year's Eve when we had been invited to a party for which I had to don my dinner-jacket. All was well until I opened a drawer in my dressing-table, when I discovered the answer to the question that had puzzled us for some time: where did the flying-squirrel store all the nuts, apple, bread and other bits of food? My brand-new cummerbund, which I had never even worn, looked like a piece of delicate Madeira lacework. The bits that had been chewed out of it had been very economically saved and used to build little nests, one on the front of each of my dress shirts. In these nests had been collected seventy-two hazel nuts, five walnuts, fourteen pieces of bread, six mealworms, fifty-two bits of apple and twenty grapes. The grapes and the apple had, of course, disintegrated somewhat with the passage of time and had left most interesting Picasso designs in juice across the front of my shirts.

I had to go to the party in a suit. The squirrel is now in Paignton Zoo.

The other day my wife said that she thought a baby otter would make a delightful pet, but I changed the subject hurriedly.

ANIMAL PARENTS

I HAVE the greatest respect for animal parents. When I was young I tried my hand at rearing a number of different creatures, and since then, on my animal-collecting trips for zoos to various parts of the world, I have had to mother quite a number of baby animals, and I have always found it a most nerve-racking task.

The first real attempt I made at being a foster-mother was to four baby hedgehogs. The female hedgehog is a very good mother. She constructs an underground nursery for the reception of her young; a circular chamber about a foot below ground-level, lined with a thick layer of dry leaves. Here she gives birth to her babies, which are blind and helpless. They are recovered with a thick coating of spikes, but these are white and soft, as though made of rubber. They gradually harden and turn brown when the babies are a few weeks old. When they are old enough to leave the nursery the mother leads them out and shows them how to hunt for food; they walk in line, rather like a school crocodile, the tail of one held in the mouth of the baby behind. The baby at the head of the column holds tight to mother's tail with grim determination, and they wend their way through the twilit hedgerows like a strange prickly centipede.

To a mother hedgehog the rearing of her babies seems to present no problems. But when I was suddenly presented with four blind, white, rubbery-spiked babies to rear, I was not so sure. We were living in Greece at the time, and the nest, which was about the size of a football and made of oak leaves, had been dug up by a peasant working in his fields. The first job was to feed the babies, for the ordinary baby's feeding-bottle only took a teat far too large for their tiny mouths. Luckily the young daughter of a friend of mine had a doll's feeding-bottle, and after much bribery I got her to part with it. After a time the hedgehogs took to this and thrived on a diet of diluted cow's milk.

I kept them at first in a shallow cardboard box where I had put the nest. But in record time the original nest was so un-hygienic that I found myself having to change the leaves ten or twelve times a day. I began to wonder if the mother hedge-hog spent her day rushing to and fro with piles of fresh leaves to keep her nest clean, and, if she did, how on earth she found time to satisfy the appetites of her babies. Mine were always ready for food at any hour of the day or night. You had only to touch the box and a chorus of shrill screams arose from four little pointed faces poking out the leaves, each head decor-ated with a crew-cut of white spikes; and the little black noses would whiffle desperately from side to side in an effort to locate the bottle.

Most baby animals know when they have had enough, but in my experience this does not apply to baby hedgehogs. Like four survivors from a raft, they flung themselves on to the bottle and sucked and sucked and sucked as though they had not had a decent meal in weeks. If I had allowed it they would have drunk twice as much as was good for them. As it was, I think I tended to overfeed them, for their tiny legs could not support the weight of their fat bodies, and they would advance across the carpet with a curious swimming motion, their tummies dragging on the ground. However, they progressed very well: their legs grew stronger, their eyes opened, and they would even make daring excursions as much as six inches away from their box.

I was very proud of my prickly family, and looked forward to the day when I would be able to take them for walks in the evening and find them delicious titbits like snails or wild strawberries. Unfortunately this dream was never realized. It so happened that I had to leave home for a day, to return the following morning. It was impossible for me to take the babies with me, so I had to leave them in charge of my sister. Before I left, I emphasized the greediness of the hedgehogs and told her that on no account were they to have more than one bottle of milk each, however much they squeaked for it.

I should have known my sister better.

When I returned the following day and inquired how my hedgehogs were, she gave me a reproachful look. I had, she said, been slowly starving the poor little things to death. With a dreadful sense of foreboding, I asked her how much she had been giving them at each meal. Four bottles each, she replied, and you should just see how lovely and fat they are getting. There was no denying they were fat. Their little tummies were so bloated their tiny feet could not even touch the ground. They looked like weird, prickly footballs to which someone by mistake had attached four legs and a nose. I did the best I could, but within twenty-four hours all four of them had died of acute enteritis. No one, of course, was more sorry than my sister, but I think she could tell by the frigid way I accepted her apologies that it was the last time she would be left in charge of any of my foster-children.

Not all animals are as good as the hedgehog at looking after their babies. Some, in fact, treat the whole business with a rather casual and modern attitude. One such is the kangaroo. Baby kangaroos are born in a very unfinished condition. They are actually embryos, for a big red kangaroo squatting on its haunches may measure five feet high and yet give birth to a baby only about half an inch long. This blind and naked blob of life has to find its way up over the mother's belly and into her pouch. In its primitive condition you would think this would be hard enough, but the whole thing is made doubly difficult by the fact that as yet the baby kangaroo can use only its front legs; the hind legs are neatly crossed over its tail. During this time the mother just squats there and gives her baby no help whatever, though occasionally she has been seen to lick a kind of trail through the fur, which may act as some sort of guide. Thus the tiny, premature offspring is forced to crawl through a jungle of fur until, as much by chance as good management, it reaches the pouch, climbs inside and clamps itself on to the teat. This is a feat that makes the ascent of Everest pale into insignificance.

I have never had the privilege of trying to hand-rear a baby kangaroo, but I have had some experience with a young wallaby, which is closely related to the species and looks just like a miniature kangaroo. I was working at Whipsnade Zoo as a keeper. The wallabies there are allowed to run free in the park, and one female, carrying a well-formed youngster, was chased by a group of young lads. In her fright she did what all the kangaroo family does in moments of stress: she tossed her youngster out of her pouch. I found it some time afterwards, lying in the long grass, twitching convulsively and making faint sucking squeaks with its mouth. It was, quite frankly, the most unprepossessing baby animal I had ever seen. About a foot long, it was blind, hairless and a bright sugar-pink. It seemed to possess no control over any part of its body except its immense hind feet, which it kicked vigorously at intervals. It had been badly bruised by its fall and I had grave doubts as to whether it would live. None the less I took it back to my lodgings and, after some argument with the landlady, kept it in my bedroom.

It fed eagerly from a bottle, but the chief difficulty lay in keeping it warm enough. I wrapped it in flannel and surrounded it with hot-water bottles, but these kept growing cold, and I was afraid it would catch a chill. The obvious thing to do was to carry it close to my body, so I put it inside my shirt. It was then that I realized for the first time what a mother wallaby must suffer. Apart from the nuzzling and sucking that went on, at regular intervals the baby would lash out its hind feet, well armed with claws, and kick me accurately in the pit of the stomach. After a few hours I began to feel as though I had been in the ring with Primo Carnera for a practice bout. It was obvious I would have to think of something else, or develop stomach ulcers. I tried putting him round the back of my shirt, but he would very soon scramble his way round to the front with his long claws in a series of convulsive kicks. Sleeping with him at night was purgatory, for apart from the all-in wrestling in which he indulged, he would sometimes kick so strongly that he shot out of bed altogether, and I was constantly forced to lean out of bed and pick him up from the floor. Unfortunately he died in two days, obviously from some sort of internal haemorrhage. I am afraid I viewed his demise with mixed feelings, although it was a pity to be deprived of the opportunity of mothering such an unusual baby.

If the kangaroo is rather dilatory about her child, the pigmy marmoset is a paragon of virtue, or rather the male is. About the size of a large mouse, clad in neat brindled green fur, and with a tiny face and bright hazel eyes, the pigmy marmoset looks like something out of a fairy tale, a small furry gnome or perhaps a kelpie. As soon as the courtship is over and the female gives birth, her diminutive spouse turns into the ideal husband. The babies, generally twins, he takes over from the moment they are born and carries them slung on his hips like a couple of saddle-bags. He keeps them clean by constant grooming, hugs them to him at night to keep them warm, and only hands them over to his rather disinterested wife at feeding time. But he is so anxious to get them back that you have the impression he would feed them himself if only he could. The pigmy marmoset is definitely a husband worth having.

Strangely enough, monkeys are generally the stupidest babies, and it takes them a long time to learn to drink out of a bottle. Having successfully induced them to do this, you have to go through the whole tedious performance again, when they are a little bit older, in an attempt to teach them to drink out of a saucer. They always seem to feel that the only way of drinking out of a saucer is to duck the face beneath the surface of the milk and stay there until you either burst for want of air or drown in your own drink.

One of the most charming baby monkeys I have ever had was a little moustached guenon. His back and tail were moss-green and his belly and whiskers a beautiful shade of buttercup yellow. Across his upper lip spread a large banana-shaped area of white, like the magnificent moustaches of some retired brigadier. Like all baby monkeys, his head seemed too big for his body, and he had long gangling limbs. He fitted very comfortably into a tea-cup. When I first had him he refused to drink out of a bottle, plainly convinced that it was some sort of fiendish torture I had invented, but eventually, when he got the hang of it, he would go quite mad when he saw the bottle arrive, fasten his mouth on to the teat, clasp the bottle passionately in his arms and roll on his back. As the bottle was at least three times his size, he made one think of a desperate survivor clinging on to a large white airship.

When he learnt, after the normal grampus-like splutterings, to drink out of a saucer, the situation became fraught with difficulty. He would be placed on a table and then his saucer of milk produced. As soon as he saw it coming he would utter a

piercing scream and start trembling all over, as if he were suffering with ague or St Vitus dance, but it was really a form of excited rage: excitement at the sight of the milk, rage that it was never put on the table quickly enough for him. He screamed and trembled to such an extent that he bounced up in the air like a grasshopper. If you were unwise enough to put the saucer down without hanging on to his tail, he would utter one final shrill scream of triumph and dive headfirst into the centre of it, and when you had mopped the resulting tidal wave of milk from your face, you would find him sitting indignantly in the middle of an empty saucer, chattering with rage because there was nothing for him to drink.

One of the main problems when you are rearing baby animals is to keep them warm enough at night, and this, strangely enough, applies even in the tropics where the temperature drops considerably after dark. In the wild state, of course, the babies cling to the dense fur of the mother and obtain warmth and shelter in that way. Hot-water bottles, as a substitute, I have found of very little use. They grow cold so quickly and you have to get up several times during the night to refill them, an exhausting process when you have a lot of baby animals to look after, as well as a whole collection of adult ones. So in most cases the simplest way is to take the babies into bed with you. You soon learn to sleep in one position – half-waking up in the night, should you wish to move, so that you avoid crushing them as you turn over.

I have at one time or another shared my bed with a great variety of young creatures, and sometimes several different species at once. On one occasion my narrow camp-bed contained three mongooses, two baby monkeys, a squirrel and a young chimpanzee. There was just enough room left over for me. You might think that after taking all this trouble a little gratitude would come your way, but in many cases you get the opposite. One of my most impressive scars was inflicted by a young mongoose because I was five minutes late with his bottle. When people ask me about it now, I am forced to pretend it was given me by a charging jaguar. Nobody would believe me if I told them it was really a baby mongoose under the bedclothes.

THE BANDITS

M Y first introduction to the extraordinary little animals known as kusimanses took place at the London Zoo. I had gone into the Rodent House to examine at close range some rather lovely squirrels from West Africa. I was just about to set out on my first animal-collecting expedition, and I felt that the more familiar I was with the creatures I was likely to meet in the great rain-forest, the easier my job would be.

After watching the squirrels for a time, I walked round the house peering into the other cages. On one of them hung a rather impressive label which informed me that the cage contained a creature known as a Kusimanse (*Crossarchus obscurus*) and that it came from West Africa. All I could see in the cage was a pile of straw that heaved gently and rhythmically, while a faint sound of snoring was wafted out to me. As I felt that this animal was one I was sure to meet, I felt justified in waking it up and forcing it to appear.

Every zoo has a rule I always observe, and many others should observe it too: not to disturb a sleeping animal by poking it or throwing peanuts. They have precious little privacy as it is. However, I ignored the rule on this occasion and rattled my thumbnail to and fro along the bars. I did not really think this would have any effect. But as I did so a sort of explosion took place in the depths of the straw, and the next moment a long, rubbery, tip-tilted nose appeared, to be followed by a rather rat-like face with small neat ears and bright inquisitive eyes. This little face appraised me for a minute; then, noticing the lump of sugar which I held tactfully near the bars, the animal uttered a faint, spinsterish squeak and struggled madly to release itself from the cocoon of straw wound round it.

When only the head had been visible, I had the impression it was only a small creature, about the size of the average ferret, but when it eventually broke loose from its covering and waddled into view, I was astonished at its relatively large body: it was, in fact, so fat as to be almost circular. Yet it shuffled over to the bars on its short legs and fell on the lump of sugar I offered, as though that was the first piece of decent food it had received in years.

It was, I decided, a species of mongoose, but its tip-tilted, whiffling nose and the glittering, almost fanatical eyes made it look totally unlike any mongoose I had ever seen. I was convinced now that its shape was due not to Nature but to overeating. It had very short legs and fine, rather slender paws, and when it trotted about the cage these legs moved so fast that they were little more than a blur beneath the bulky body. Each time I fed it a morsel of food it gave the same faint, breathless squeak: as much as to reproach me for tempting it away from its diet.

I was so captivated by this little animal that before I realized what I was doing I had fed it all the lump-sugar in my pocket. As soon as it knew that no more titbits were forthcoming, it uttered a long-suffering sigh and trotted away to dive into the straw. Within a couple of seconds it was sound asleep once more. I decided there and then that if kusimanses were to be obtained in the area I was visiting, I would strain every nerve to find one.

Three months later I was deep in the heart of the Cameroon rain-forests and here I found I had ample opportunity for getting to know the kusimanse. Indeed, they were about the commonest members of the mongoose family, and I often saw them when I was sitting concealed in the forest waiting for some completely different animal to make its appearance.

The first one I saw appeared suddenly out of the under-growth on the banks of a small stream. He kept me amused for a long time with a display of his crab-catching methods: he waded into the shallow water and with the aid of his long, turned-up nose (presumably holding his breath when he did so) he turned over all the rocks he could find until he un-earthed one of the large, black, freshwater crabs. Without a second's hesitation he grabbed it in his mouth and, with a quick flick of his head, tossed it on to the bank. He then chased after it, squeaking with delight and danced round it, snapping away until at last it was dead. When an exception-ally large crab succeeded in giving him a nip on the end of his *retroussé* nose, I am afraid my stifled amusement caused the kusimanse to depart hastily into the forest.

On another occasion I watched one of these little beasts using precisely the same methods to catch frogs, but this time without much success. I felt he must be young and inexperi-enced in the art of frog-catching. After much laborious hunt-ing and snuffling, he would catch a frog and hurl it shorewards; but, long before he had waddled out to the bank after it, the frog would have recovered itself and leapt back into the water, and the kusimanse would be forced to start all over again.

One morning a native hunter walked into my camp carry-ing a small palm-leaf basket, and peering into it I saw three of the strangest little animals imaginable. They were about the size of new-born kittens, with tiny legs and somewhat moth-eaten tails. They were covered with bright gingery-red fur which stood up in spikes and tufts all over their bodies, making them look almost like some weird species of hedgehog. As I gazed down at them, trying to identify them, they lifted their little faces and peered up at me. The moment I saw the long,

pink, rubbery noses I knew they were kusimanses, and very young ones at that, for their eyes were only just open and they had no teeth. I was very pleased to obtain these babies, but after I had paid the hunter and set to work on the task of trying to teach them to feed, I began to wonder if I had not got more than I bargained for. Among the numerous feeding-bottles I had brought with me I could not find a teat small enough to fit their mouths, so I was forced to try the old trick of wrapping some cotton-wool round the end of a matchstick, dipping it in milk and letting them suck it. At first they took the view that I was some sort of monster endeavouring to choke them. They struggled and squeaked, and every time I pushed the cotton-wool into their mouths they frantically spat it out again. Fortunately it was not long before they discovered that the cotton-wool contained milk, and then they were no more trouble, except that they were liable to suck so hard in their enthusiasm that the cotton-wool would part company with the end of the matchstick and disappear down their throats.

At first I kept them in a small basket by my bed. This was the most convenient spot, for I had to get up in the middle of the night to feed them. For the first week or so they really behaved very well, spending most of the day sprawled on their bed of dried leaves, their stomachs bulging and their paws twitching. Only at meal-times would they grow excited, scrambling round and round inside the basket, uttering loud squeaks and treading heavily on one another.

It was not long before the baby kusimanses developed their front teeth (which gave them a firmer and more disastrous grip on the cotton-wool), and as their legs got stronger they became more and more eager to see the world that lay outside their basket. They had the first feed of the day when I drank my morning tea; and I would lift them out of their basket and put them on my bed so that they could have a walk round. I had, however, to call an abrupt halt to this habit, for one morning, while I was quietly sipping my tea, one of the baby kusimanses discovered my bare foot sticking out from under the bedclothes and decided that if he bit my toe hard enough it might produce milk. He laid hold with his needle-sharp teeth, and his brothers, thinking they were missing a feed, instantly joined him. When I had locked them up in their basket again and finished mopping tea off myself and the bed, I decided these morning romps would have to cease. They were too painful.

This was merely the first indication of the trouble in store for me. Very soon they had become such a nuisance that I was forced to christen them the Bandits. They grew fast, and as soon as their teeth had come through they started to eat egg and a little raw meat every day, as well as their milk. Their appetites seemed insatiable, and their lives turned into one long quest for food. They appeared to think that everything was edible unless proved otherwise. One of the things of which they made a light snack was the lid of their basket. Having demolished this they hauled themselves out and went on a tour of inspection round the camp. Unfortunately, and with unerring accuracy, they made their way to the one place where they could do the maximum damage in the minimum time: the place where the food and medical supplies were stored. Before I discovered them they had broken a dozen eggs and, to judge by the state of them, rolled in the contents. They had fought with a couple of bunches of bananas and apparently won, for the bananas looked distinctly the worse for wear. Having slaughtered the fruit, they had moved on and upset two bottles of vitamin product. Then, to their delight, they had found two large packets of boracic powder. These they had burst open and scattered far and wide, while large quantities of the white powder had stuck to their egg-soaked fur. By the time I found them they were on the point of having a quick drink from a highly pungent and poisonous bucket of disinfectant, and I grabbed them only just in time. Each of them looked like some weird Christmas cake decoration, in a coat stiff with boracic and egg yolk. It took me three quarters of an hour to clean them up. Then I put them in a larger and stronger basket and hoped that this would settle them.

It took them two days to break out of *this* basket.

This time they had decided to pay a visit to all the other animals I had. They must have had a fine time round the cages, for there were always some scraps of food lying about.

Now at that time I had a large and very beautiful monkey, called Colly, in my collection. Colly was a colobus, perhaps one of the most handsome of African monkeys. Their fur is coal black and snow white, hanging in long silky strands round their bodies like a shawl. They have a very long plume-like tail, also black and white. Colly was a somewhat vain monkey and spent a lot of her time grooming her lovely coat and posing in various parts of the cage. On this particular afternoon she had decided to enjoy a siesta in the bottom of her box, while waiting for me to bring her some fruit. She lay there like a sunbather on a beach, her eyes closed, her hands folded neatly on her chest. Unfortunately, however, she had pushed her tail through the bars so that it lay on the ground outside like a feathery black-and-white scarf that someone had dropped. Just as Colly was drifting off into a deep sleep, the Bandits appeared on the scene.

The Bandits, as I pointed out, believed that everything in the world, no matter how curious it looked, might turn out to be edible. In their opinion it was always worth sampling everything, just in case. When he saw Colly's tail lying on the ground ahead, apparently not belonging to anyone, the eldest Bandit decided it must be a tasty morsel of something or other that Providence had placed in his path. So he rushed forward and sank his sharp little teeth into it. His two brothers, feeling that there was plenty of this meal for everyone, joined him immediately. Thus was Colly woken out of a deep and refreshing sleep by three sets of extremely sharp little teeth fastening themselves almost simultaneously in her tail. She gave a wild scream of fright and scrambled towards the top of her cage. But the Bandits were not going to be deprived of this tasty morsel without a struggle, and they hung on grimly. The higher Colly climbed in her cage, the higher she lifted the Bandits off the ground, and when eventually I got there in response to her yells, I found the Bandits, like some miniature trapeze-artists, hanging by their teeth three feet off the ground. It took me five minutes to make them let go, and then I managed it only by blowing cigarette-smoke in their faces and making them sneeze. By the time I had got them safely locked up again, poor Colly was a nervous wreck.

I decided the Bandits must have a proper cage if I did not want the rest of my animals driven hysterical by their attentions. I built them a very nice one, with every modern convenience. It had a large and spacious bedroom at one end, and an open playground and dining-room at the other. There were two doors, one to admit my hand to their bedroom, the other to put their food into their dining-room. The trouble lay in feeding them. As soon as they saw me approach with a plate they would cluster round the doorway, screaming excitedly, and the moment the door was opened they would shoot out, knock the plate from my hand and fall to the ground with it, a tangled mass of kusimanses, raw meat, raw egg and milk. Quite often when I went to pick them up they would bite me, not vindictively but simply because they would mistake my fingers for something edible. Yes, feeding the Bandits was not only a wasteful process but an extremely painful one as well. By the time I got them safely back to England they had bitten me twice as frequently as any animal I have ever kept. So it was with a real feeling of relief that I handed them over to a zoo.

The next day I went round to see how they were settling down. I found them in a huge cage, pattering about and looking, I felt, rather lost and bewildered by all the new sights and sounds. Poor little things, I thought, they have had the wind taken out of their sails. They looked so subdued and forlorn. I began to feel quite sorry to have parted with them. I stuck my finger through the wire and waggled it, calling to them. I thought it might comfort them to talk to someone they knew. I should have known better: the Bandits shot across the cage in a grim-faced bunch and fastened on to my finger like bulldogs. With a yelp of pain I at last managed to get my finger away, and as I left them, mopping the blood from my hand, I decided that perhaps, after all, I was not *so* sorry to see the back of them. Life without the Bandits might be considerably less exciting – but it would not hurt nearly so much.

WILHELMINA

Most people, when they learn for the first time that I collect wild animals for zoos, ask the same series of questions in the same order. First they ask if it is dangerous, to which the answer is no, it is not, providing you do not make any silly mistakes. Then they ask how I catch the animals – a more difficult question to answer, for there are many hundreds of ways of capturing wild animals: sometimes you have no set method, but have to improvise something on the spur of the moment. Their third question is, invariably: don't you become attached to your animals and find it difficult to part with them at the end of an expedition? The answer is, of course, that you do, and sometimes parting with a creature you have kept for eight months can be a heartbreaking process.

Occasionally you even find yourself getting attached to the strangest of beasts, some weird creature you would never in the normal way have thought you could like. One such beast as this, I remember, was Wilhelmina.

Wilhelmina was a whip-scorpion, and if anyone had told me that the day would come when I would feel even the remotest trace of affection for a whip-scorpion I would never have believed them. Of all the creatures on the face of this earth the whip-scorpion is one of the least prepossessing. To those who do not adore spiders (and I am one of those people) the whip-scorpion is a form of living nightmare. It resembles a spider with a body the size of a walnut that has been run over by a steamroller and flattened to a wafer-thin flake. To this flake are attached what appear to be an immense number of long, fine and crooked legs which spread out to the size of a soup-plate. To cap it all, on the front (if such a creature can be said to have a front), are two enormously long slender legs like whips, about twelve inches long in a robust specimen. It possesses the ability to skim about at incredible speed and with apparently no effort – up, down or sideways – and to squeeze its revolting body into a crack that would scarcely accommodate a piece of tissue-paper.

That is a whip-scorpion, and to anyone who distrusts spiders it is the personification of the devil. Fortunately they are harmless, unless you happen to have a weak heart.

I made my first acquaintance with Wilhelmina's family when I was on a collecting trip to the tropical forest of West Africa. For many different reasons, hunting in these forests is always difficult. To begin with, the trees are enormous, some as much as a hundred and fifty feet high, with trunks as fat as a factory chimney. Their head foliage is thick, luxuriant and twined with creepers and the branches are decorated with various parasitic plants like a curious hanging garden. All this may be eighty or a hundred feet above the forest floor, and the only way to reach it is to climb a trunk as smooth as a plank which has not a single branch for the first seventy feet of its length. This, the top layer of the forest, is by far the most thickly populated, for in the comparative safety of the tree-tops live a host of creatures which rarely, if ever, descends to ground-level. Setting traps in the forest canopy is a difficult and tedious operation. It may take a whole morning to find a way up a tree, climb it and set the trap in a suitable position. Then, just as you have safely regained the forest floor, your trap goes off with a triumphant clang, and the whole laborious process has to be endured once more. Thus, although trap-setting in the tree-tops is a painful necessity, you are always on the look-out for some slightly easier method of obtaining the animals you want. Probably one of the most successful and exciting of these methods is to smoke out the giant trees.

Some of the forest trees, although apparently sound and solid, are actually hollow for part or all of their length. These are the trees to look for, though they are not so easy to find. A day of searching in the forest might end with the discovery of six of them, perhaps one of which will yield good results when finally smoked out.

Smoking out a hollow tree is quite an art. To begin with, you must, if necessary, enlarge the opening at the base of the trunk and lay a small fire of dry twigs. Then two Africans are sent up the tree with nets to cover all the holes and cracks at the upper end of the trunk, and then station themselves at convenient points to catch any animals that emerge. When all is ready, you start the fire, and as soon as it is crackling you lay on top of the flames a large bundle of fresh green leaves. Immediately the flames die away and in their place rises a column of thick and pungent smoke. The great hollow interior of the tree acts like a gigantic chimney, and the smoke is whisked up inside. You never realize, until you light the fire, quite how many holes and cracks there are in the trunk of the tree. As you watch, you see a tiny tendril of smoke appear magically on the bark perhaps twenty feet from the ground, coiling out of an almost invisible hole; a short pause and ten feet higher three more little holes puff smoke like miniature cannon-mouths. Thus, guided by the tiny streamers of smoke appearing at intervals along the trunk, you can watch the progress of the smoking. If the tree is a good one, you have only time to watch the smoke get half-way up, for it is then that the animals start to break cover and you become very busy indeed.

When one of these hollow trees is inhabited, it is really like a block of flats. In the ground-floor apartments, for example, you find things like the giant land-snails, each the size of an apple, and they come gliding out of the base of the tree with all the speed a snail is capable of mustering, even in an emergency. They may be followed by other creatures who prefer the lower apartments or else are unable to climb: the big forest toads, for example, whose backs are cleverly marked out to resemble a dead leaf, and whose cheeks and sides are a beautiful mahogany red. They come waddling out from among the tree-tops with the most ludicrously indignant expressions on their faces, and on reaching the open air suddenly squat down and stare about them in a pathetic and helpless sort of way.

Having evicted all the ground-floor tenants, you then have to wait a short time before the occupants higher up have a chance to make their way down to the opening. Almost invariably giant millipedes are among the first to appear – charming creatures that look like brown sausages, with a fringe of legs along the underside of their bodies. They are quite harmless and rather imbecile creatures for which I have a very soft spot. One of their most ridiculous antics, when placed on a table, is to set off walking, all their legs working furiously, and on coming to the edge they never seem to notice it and continue to walk out into space until the weight of their body bends them over. Then, half on and half off the table, they pause, consider, and eventually decide that something is wrong. And so, starting with the extreme hind pair of legs, they go into reverse and get themselves on to the table again – only to crawl to the other side and repeat the performance.

Immediately after the appearance of the giant millipedes all the other top-floor tenants of the tree break cover together, some making for the top of the tree, others for the bottom. Perhaps there are squirrels with black ears, green bodies and tails of the most beautiful flame colour; giant grey dormice who gallop out of the tree, trailing their bushy tails behind them like puffs of smoke; perhaps a pair of bush-babies, with their great innocent eyes and their slender attenuated and trembly hands, like those of very old men. And then, of course, there are the bats: great fat brown bats with curious flower-like decorations on the skin of their noses and large transparent ears; others bright ginger, with black ears twisted down over their heads and pig-like snouts. And as this pageant of wild life appears the whip-scorpions are all over the place, skimming up and down the tree with a speed and silence that is un-nerving and uncanny, squeezing their revolting bodies into the thinnest crack as you make a swipe at them with the net, only to reappear suddenly ten feet lower down the tree, skimming towards you apparently with the intention of disappear-ing into your shirt. You step back hurriedly and the creature vanishes: only the tips of a pair of antennae, wiggling from the depths of a crevice in the bark that would hardly accom-modate a visiting-card, tells you of its whereabouts. Of the many creatures in the West African forest the whip-scorpion has been responsible for more shocks to my system than any other. The day a particularly large and leggy specimen ran over my bare arm, as I leant against a tree, will always be one of my most vivid memories. It took at least a year off my life.

But to return to Wilhelmina. She was a well-brought-up little whip-scorpion, one of a family of ten, and I started my intimate acquaintance with her when I captured her mother. All this happened quite by chance.

I had for many days been smoking out trees in the forest in search of an elusive and rare little animal known as the pigmy scaly-tail. These little mammals, which look like mice with long feathery tails, have a curious membrane of skin stretched from ankle to wrist, with the aid of which they glide around the forest with the ease of swallows. The scaly-tails live in colonies in hollow trees, but the difficulty lay in finding a tree that contained a colony. When, after much fruitless hunting, I did discover a group of these prizes, and moreover actually managed to capture some, I felt considerably elated. I even started to take a benign interest in the numerous whip-scorpions that were scuttling about the tree. Then suddenly I noticed one which looked so extraordinary, and was behaving in such a peculiar manner, that my attention was at once arrested. To begin with, this whip-scorpion seemed to be wearing a green fur-coat that almost completely covered her chocolate body. Secondly, it was working its way slowly and carefully down the tree with none of the sudden fits and starts common to the normal whip-scorpion.

Wondering if the green fur-coat and the slow walk were symptoms of extreme age in the whip-scorpion world I moved closer to examine the creature. To my astonishment I found that the fur-coat was composed of baby whip-scorpions, each not much larger than my thumb-nail, which were obviously fairly recent additions to the family. They were, in extra-ordinary contrast to their dark-coloured mother, a bright and bilious green, the sort of green that confectioners are fond of using in cake decorations. The mother's slow and stately progress was due to her concern lest one of her babies lose its grip and drop off. I realized, rather ruefully, that I had never given the private life of the whip-scorpion much thought, and it had certainly never occurred to me that the female would be sufficiently maternal to carry her babies on her back. Over-come with remorse at my thoughtlessness, I decided that here was an ideal chance for me to catch up on my studies of these creatures. So I captured the female very carefully – to avoid dropping any of her progeny – and carried her back to camp.

I placed the mother and children in a large roomy box with plenty of cover in the way of bark and leaves. Every morning I had to look under these, rather gingerly I admit, to see if she was all right. At first, the moment I lifted the bark under which she was hiding, she would rush out and scuttle up the side of the box, a distressing habit which always made me jump and slam the lid down. I was very much afraid that one day I might do this and trap her legs or antennae, but fortunately after the first three days or so she settled down, and would even let me renew the leaves and bark in her box without taking any notice.

I had the female whip-scorpion and her babies for two months, and during that time the babies ceased to ride on their mother's back. They scattered and took up residence in various parts of the box, grew steadily and lost their green colouring in favour of brown. Whenever they grew too big for their skins they would split them down the back and step out of them, like spiders. Each time they did so they would emerge a little larger and a little browner. I discovered that while the mother would tackle anything from a small grasshopper to a large beetle, the babies were fussy and demanded small spiders, slugs and other easily digestible fare. They all appeared to be thriving, and I began to feel rather proud of them. Then one day I returned to camp after a few hours hunting in the forest to find that tragedy had struck.

A tame Patas monkey I kept tied up outside the tent had eaten through his rope and been on a tour of investigation. Before anyone had noticed it he had eaten a bunch of bananas, three mangos and four hard-boiled eggs, he had broken two bottles of disinfectant, and rounded the whole thing off by knocking my whip-scorpion box on to the floor. It promptly broke open and scattered the family on the ground, and the Patas monkey, a creature of depraved habits, had set to work and eaten them. When I got back he was safely tied up again, and suffering from an acute attack of hiccups.

I picked up my whip-scorpion nursery and peered mournfully into it, cursing myself for having left it in such an accessible place, and cursing the monkey for having such an

appetite. But then, to my surprise and delight, I found, squatting in solitary state on a piece of bark, one of the baby whip-scorpions, the sole survivor of the massacre. Tenderly I moved it to a smaller and more burglar-proof cage, showered it with slugs and other delicacies and christened it, for no reason at all, Wilhelmina.

During the time I had Wilhelmina's mother, and Wilhelmina herself, I learnt quite a lot about whip-scorpions. I discovered that though quite willing to hunt by day if hungry, they were at their most lively during the night. During the day Wilhelmina was always a little dull-witted, but in the evening she woke up and, if I may use the expression, blossomed. She would stalk to and fro in her box, her pincers at the ready, her long antennae-like legs lashing out like whips ahead of her, seeking the best route. Although these tremendously elongated legs are supposed to be merely feelers, I got the impression that they could do more than this. I have seen them wave in the direction of an insect, pause and twitch, whereupon Wilhelmina would brace herself, almost as if she had smelt or heard her prey with the aid of her long legs. Sometimes she would stalk her food like this; at other times she would simply lie in wait until the unfortunate insect walked almost into her arms, and the powerful pincers would gather it lovingly into her mouth.

As she grew older I gave her bigger and bigger things to eat, and I found her courage extraordinary. She was rather like a pugnacious terrier who, the larger the opponent, the better he likes the fight. I was so fascinated by her skill and bravery in tackling insects as big or bigger than herself that one day, rather unwisely, I put a very large locust in with her. Without a moment's hesitation, she flew at him and grasped his bulky body in her pincers. To my alarm, however, the locust gave a hearty kick with his powerful hind legs and both he and

Wilhelmina soared upwards and hit the wire-gauze roof of the cage with a resounding thump, then crashed back to the floor again. This rough treatment did not deter Wilhelmina at all, and she continued to hug the locust while he leapt wildly around the cage, thumping against the roof, until eventually he was exhausted. Then she settled down and made short work of him. But after this I was always careful to give her the smaller insects, for I had visions of a leg or one of her whips being broken off in such a rough contest.

By now I had become very fond and not a little proud of Wilhelmina. She was, as far as I knew, the only whip-scorpion to have been kept in captivity. What is more, she had become very tame. I had only to rap on the side of her box with my fingers and she would appear from under her piece of bark and wave her whips at me. Then, if I put my hand inside, she would climb on to my palm and sit there quietly while I fed her with slugs, creatures for which she still retained a passion.

When the time drew near for me to transport my large collection of animals back to England, I began to grow rather worried over Wilhelmina. It was a two-week voyage, and I could not take enough insect food for that length of time. I decided therefore to try making her eat raw meat. It took me a long time to achieve it, but once I had learnt the art of waggling the bit of meat seductively enough I found that Wilhelmina would grab it, and on this unlikely diet she seemed to thrive. On the journey down to the coast by lorry Wilhelmina behaved like a veteran traveller, sitting in her box and sucking a large chunk of raw meat almost throughout the trip. For the first day on board ship the strange surroundings made her a little sulky, but after that the sea air seemed to do her good and she became positively skittish. This was her undoing.

One evening when I went to feed her, she scuttled up as far as my elbow before I knew what was happening, dropped on to a hatch-cover and was just about to squeeze her way through a crack on a tour of investigation when I recovered from my astonishment and managed to grab her. For the next few days I fed her very cautiously, and she seemed to have quietened down and regained her former self-possession.

Then one evening she waggled her whips at me so plaintively that I lifted her out of her cage on the palm of my hand and started to feed her on the few remaining slugs I had brought for her in a tin. She ate two slugs, sitting quietly and decorously on my hand, and then suddenly she jumped. She could not have chosen a worse time, for as she was in mid-air a puff of wind swept round the bulkhead and whisked her away. I had a brief glimpse of her whips waving wildly, and then she was over the rail and gone, into the vast heaving landscape of the sea. I rushed to the rail and peered over, but it was impossible to spot so small a creature in the waves and froth below. Hurriedly I threw her box over, in the vain hope that she might find it and use it as a raft. A ridiculous hope, I know, but I did not like to think of her drowning without making some attempt to save her. I could have kicked myself for my stupidity in lifting her out of her box; I never thought I would have been so affected by the loss of such a creature. I had grown very fond of her; she in her turn had seemed to trust me. It was a tragic way for the relationship to end. But there was one slight consolation: after my association with Wilhelmina I shall never again look at a whip-scorpion with quite the same distaste.

ADOPTING AN ANTEATER

Making a collection of two hundred birds, mammals and reptiles is rather like having two hundred delicate babies to look after. It needs a lot of hard work and patience. You have to make sure their diet suits them, that their cages are big enough, that they get neither too hot in the tropics nor too cold when you get near England. You have to de-worm, detick and de-flea them; you have to keep their cages and feeding-pots spotlessly clean.

But, above all, you have to make sure that your animals are *happy*. However well looked after, a wild animal will not live in captivity unless it is happy. I am talking, of course, of the adult, wild-caught creature. But occasionally you get a baby wild animal whose mother has perhaps met with an accident, and who has been found wandering in the forest. When you capture one of these, you must be prepared for a good deal of hard work and worry, and above all you must be ready to give the animal the affection and confidence it requires; for after a day or two you will have become the parent, and the baby will trust you and depend on you completely.

This can sometimes make life rather difficult. There have been periods when I have played the adopted parent to as many as six baby animals at once, and this is no joke. Quite apart from anything else, imagine rising at three o'clock in the morning, stumbling about, half-asleep, in an effort to prepare six different bottles of milk, trying to keep your eyes open enough to put the right amount of vitamin drops and sugar in, knowing all the time that you will have to be up again in three hours to repeat the performance.

Some time ago my wife and I were on a collecting trip in

Paraguay, that country shaped like a boot-box which lies almost in the exact centre of South America. Here, in a remote part of Chaco, we assembled a lovely collection of animals. Many things quite unconnected with animals happen on a collecting trip, things that frustrate your plans or irritate you in other ways. But politics, mercifully, had never before been among them. On this occasion, however, the Paraguayans decided to have a revolution, and as a direct result we had to release nearly the whole of our collection and escape to Argentina in a tiny four-seater plane.

Just before our retreat, an Indian had wandered into our camp carrying a sack from which had come the most extraordinary noises. It sounded like a cross between a cello in pain and a donkey with laryngitis. Opening the sack, the Indian tipped out one of the most delightful baby animals I had ever seen. She was a young giant anteater, and she could not have been more than a week old. She was about the size of a corgie, with black, ash-grey and white fur, a long slender snout and a pair of tiny, rather bleary eyes. The Indian said he had found her wandering about in the forest, honking forlornly. He thought her mother might have been killed by a jaguar.

The arrival of this baby put me in a predicament. I knew that we would be leaving soon and that the plane was so tiny that most of our equipment would have to be left behind to make room for the five or six creatures we were determined to take with us. To accept, at that stage, a baby anteater who weighed a considerable amount and who would have to be fussed over and bottle-fed, would be lunatic. Quite apart from anything else, no one as far as I knew, had ever tried to rear a baby anteater on a bottle. The whole thing was obviously out of the question. Just as I had made up my mind the baby, still blaring pathetically, suddenly discovered my leg, and with a honk of joy shinned up it, settled herself in my

lap and went to sleep. Silently I paid the Indian the price he demanded, and thus became a father to one of the most charming children I have ever met.

The first difficulty cropped up almost at once. We had a baby's feeding-bottle, but we had exhausted our supply of teats. Luckily a frantic house-to-house search of the little village where we were living resulted in the discovery of one teat, of extreme age and unhygienic appearance. After one or two false starts the baby took to the bottle far better than I had dared hope, though feeding her was a painful performance.

Young anteaters, at that age, cling to their mother's back, and, since we had, so to speak, become her parents, she insisted on climbing on to one or the other of us nearly the whole time. Her claws were about three inches long, and she had a prodigious grip with them. During meals she clasped your leg affectionately with three paws, while with her remaining paw held your finger and squeezed it hard at intervals, for she was convinced that this would increase the flow of milk from the bottle. At the end of each feed you felt as though you had been mauled by a grizzly bear, while your fingers had been jammed in a door.

For the first days I carried her about with me to give her confidence. She liked to lie across the back of my neck, her long nose hanging down one side of me and her long tail down the other, like a fur collar. Every time I moved she would tighten her grip in a panic, and this was painful. After the fourth shirt had been ruined I decided that she would have to cling to something else, so I filled a sack full of straw and introduced her to that. She accepted it without any fuss, and so between meals she would lie in her cage, clutching this substitute happily. We had already christened her 'Sarah', and now that she developed this habit of sack-clutching we gave her a surname, and so she became known as 'Sarah Huggersack'.

Sarah was a model baby. Between feeds she lay quietly on her sack, occasionally yawning and showing a sticky, pinky-grey tongue about twelve inches long. When feeding-time came round she would suck the teat on her bottle so vigorously that it had soon changed from red to pale pink, the hole at the end of it had become about the size of a matchstick, and the whole thing drooped dismally from the neck of the bottle.

When we had to leave Paraguay in our extremely unsafe-looking four-seater plane, Sarah slept peacefully throughout the flight, lying on my wife's lap and snoring gently, occasionally blowing a few bubbles of sticky saliva out of her nose.

On arriving in Buenos Aires our first thought was to give Sarah a treat. We would buy her a nice new shiny teat. We went to endless trouble selecting one exactly the right size, shape and colour, put it on the bottle and presented it to Sarah. She was scandalized. She honked wildly at the mere thought of a new teat, and sent the bottle flying with a well-directed clout from her paw. Nor did she calm down and start to feed until we had replaced the old withered teat on the bottle. She clung to it ever after; months after her arrival in England she still refused to be parted from it.

In Buenos Aires we housed our animals in an empty house on the outskirts of the city. From the centre, where we stayed, it took us half an hour in a taxi to reach it, and this journey we had to do twice and sometimes three times a day. We soon found that having a baby anteater made our social life difficult in the extreme. Have you ever tried to explain to a hostess that you must suddenly leave in the middle of dinner because you have to give a bottle to an anteater? In the end our friends gave up in despair. They used to telephone and ascertain the times of Sarah's feeds before inviting us.

By this time Sarah had become much more grown up and independent. After her evening feed she would go for a walk round the room by herself. This was a great advance, for up till then she had screamed blue murder if you moved more than a foot or so away from her. After her tour of inspection she liked to have a game. This consisted in walking past us, her nose in the air, her tail trailing temptingly. You were then supposed to grab the end of her tail and pull, whereupon she would swing round on three legs and give you a gentle clout with her paw. When this had been repeated twenty or thirty times she felt satisfied, and then you had to lay her on her back and tickle her tummy for ten minutes or so while she closed her eyes and blew bubbles of ecstasy at you. After this she would go to bed without any fuss. But try to put her to bed without giving her a game and she would kick and struggle and honk, and generally behave in a thoroughly spoilt manner.

When we eventually got on board ship, Sarah was not at all sure that she approved of sea-voyages. To begin with, the ship smelt queer; then there was a strong wind which nearly blew her over every time she went for a walk on deck; and lastly, which she hated most of all, the deck would not keep still. First it tilted one way, then it tilted another, and Sarah would go staggering about, honking plaintively, banging her nose on bulkheads and hatch-covers. When the weather improved, however, she seemed to enjoy the trip. Sometimes in the afternoon, when I had time, I would take her up to the promenade deck and we would sit in a deck-chair and sunbathe. She even paid a visit to the bridge, by special request of the captain. I thought it was because he had fallen for her charm and personality, but he confessed that it was because (having seen her only from a distance) he wanted to make sure which end of her was the front.

I must say we felt very proud of Sarah when we arrived in

London Docks and she posed for the Press photographers with all the unselfconscious ease of a born celebrity. She even went so far as to lick one of the reporters – a great honour. I hastily tried to point this out to him, while helping to remove a large patch of sticky saliva from his coat. It was not everyone she would lick, I told him. His expression told me that he did not appreciate the point.

Sarah went straight from the docks to a zoo in Devonshire, and we hated to see her go. However, we were kept informed about her progress and she seemed to be doing well. She had formed a deep attachment to her keeper.

Some weeks later I was giving a lecture at the Festival Hall, and the organizer thought it would be rather a good idea if I introduced some animal on the stage at the end of my talk. I immediately thought of Sarah. Both the zoo authorities and the Festival Hall Management were willing, but, as it was now winter, I insisted that Sarah must have a dressing-room to wait in.

I met Sarah and her keeper at Paddington Station. Sarah was in a huge crate, for she had grown as big as a red setter, and she created quite a sensation on the platform. As soon as she heard my voice she flung herself at the bars of her cage and protruded twelve inches of sticky tongue in a moist and affectionate greeting. People standing near the cage leapt back hurriedly, thinking some curious form of snake was escaping and it took a lot of persuasion before we could find a porter brave enough to wheel the cage on a truck.

When we reached the Festival Hall we found that the rehearsal of a symphony concert had just come to an end. We wheeled Sarah's big box down long corridors to the dressing-room, and just as we reached the door it was flung open and Sir Thomas Beecham strode out, smoking a large cigar. We wheeled Sarah into the dressing-room he had just vacated.

While I was on the stage, my wife kept Sarah occupied by running round and round the dressing-room with her, to the consternation and horror of one of the porters, who, hearing the noise, was convinced that Sarah had broken out of her cage and was attacking my wife. Eventually, however, the great moment arrived and amid tumultuous applause Sarah was carried on to the stage. She was very short-sighted, as all ant-eaters are, so to her the audience was non-existent. She looked round vaguely to see where the noise was coming from, but decided that it was not really worth worrying about. While I extolled her virtues, she wandered about the stage, oblivious, occasionally snuffling loudly in a corner, and repeatedly approaching the microphone and giving it a quick lick, which left it in a very sticky condition for the next performer. Just as I was telling the audience how well-behaved she was, she discovered the table in the middle of the stage, and with an immense sigh of satisfaction proceeded to scratch her bottom against one of the legs. She was a great success.

After the show, Sarah held court for a few select guests in her dressing-room, and became so skittish that she even galloped up and down outside in the corridor. Then we bundled her up warmly and put her on the night train for Devon with her keeper.

Apparently, on reaching the zoo again, Sarah was thoroughly spoilt. Her short spell as a celebrity had gone to her head. For three days she refused to be left alone, stamping about her cage and honking wildly, and refusing all food unless she was fed by hand.

A few months later I wanted Sarah to make an appearance on a television show I was doing, and so once again she tasted the glamour and glitter of show business. She behaved with the utmost decorum during rehearsals, except that she was dying to investigate the camera closely, and had to be restrained by force. When the show was over she resisted going back to her cage, and it took the united efforts of myself, my wife, Sarah's keeper and the studio manager to get her back into the box – for Sarah was then quite grown up, measuring six feet from nose to tail, standing three feet at the shoulder and with forearms as thick as my thigh.

We did not see Sarah again until quite recently, when we paid her a visit at her zoo. It had been six months since she had last seen us, and quite frankly I thought she would have forgotten us. Anteater fan though I am, I would be the first to admit that they are not creatures who are overburdened with brains, and six months is a long time. But the moment we called to her she came bounding out of her sleeping den and rushed to the wire to lick us. We even went into the cage and played with her, a sure sign that she really did recognize us, for no one else except her keeper dared enter.

Eventually we said good-bye to her, rather sadly, and left her sitting in the straw blowing bubbles after us. As my wife said: 'It was rather as though we were leaving our child at boarding school.' We are certainly her adopted parents, as far as Sarah is concerned.

Yesterday we had some good news. We heard that Sarah has got a mate. He is as yet too young to be put in with her, but soon he should be big enough. Who knows, by this time next year we may be grandparents to a fine bouncing baby anteater!

PORTRAIT OF PAVLO

IT is a curious thing, but when you keep animals as pets you tend to look upon them so much as miniature human beings that you generally manage to impress some of your own characteristics on to them. This anthropomorphic attitude is awfully difficult to avoid. If you possess a golden hamster and are always watching the way he sits up and eats a nut, his little pink paws trembling with excitement, his pouches bulging as he saves in his cheeks what cannot be eaten immediately, you might one day come to the conclusion that he looks exactly like your own Uncle Amos sitting, full of port and nuts, in his favourite club. From that moment the damage is done. The hamster continues to behave like a hamster, but you regard him only as a miniature Uncle Amos, clad in a ginger fur-coat, for ever sitting in his club, his cheeks bulging with food. There are very few animals who have characters strong and distinct enough to overcome this treatment, who display such powerful personalities that you are forced to treat them as individuals and not as miniature human beings. Of the many hundreds of animals I have collected for zoos in this country, and of the many I have kept as pets, I can remember at the most about a dozen creatures who had this strength of personality that not only made them completely different from others of their kind, but enabled them to resist all attempts on my part to turn them into something they were not.

One of the smallest of these animals was Pavlo, a black-eared marmoset, and his story really started one evening when, on a collecting trip in British Guiana, I sat quietly in the bushes near a clearing, watching a hole in a bank which I had good reason to believe contained an animal of some description. The sun was setting and the sky was a glorious salmon pink, and outlined against it were the massive trees of the forest, their branches so entwined with creepers that each tree looked as though it had been caught in a giant spider's web. There is nothing quite so soothing as a tropical forest at this time of day. I sat there absorbing sights and colours, my mind in the blank and receptive state that the Buddhists tell us is the first step towards Nirvana. Suddenly my trance was shattered by a shrill and prolonged squeak of such intensity that it felt as though someone had driven a needle into my ear. Peering above me cautiously, I tried to see where the sound had come from: it seemed the wrong sort of note for a tree-frog or an insect, and far too sharp and tuneless to be a bird. There, on a great branch about thirty feet above me, I saw the source of the noise: a diminutive marmoset was trotting along a wide branch as if it were an arterial road, picking his way in and out of the forest of orchids and other parasitic plants that grew in clumps from the bark. As I watched, he stopped, sat up on his hind legs and uttered another of his piercing cries; this time he was answered from some distance away, and within a moment or two other marmosets had joined him. Trilling and squeaking to each other, they moved among the orchids, searching diligently, occasionally uttering shrill squeaks of joy as they unearthed a cockroach or a beetle among the leaves. One of them pursued something through an orchid plant for a long time, parting the leaves and peering between them with an intense expression on his tiny face. Every time he made a grab the leaves got in the way and the insect managed to escape round the other side of the plant. Eventually, more by good luck than skill, he dived his small hands in amongst the leaves and, with a twitter of triumph, emerged with a fat cockroach clutched firmly between his fingers. The insect was a large one and its wriggling was strenuous, so, presumably in case he dropped it, he stuffed the whole thing into his mouth. He sat there munching happily, and when he had swallowed the last morsel, he carefully examined both the palms and backs of his paws to make sure there was none left.

I was so entranced by this glimpse into the private life of the marmoset that it was not until the little party had moved off into the now-gloomy forest that I realized I had an acute crick in my neck and that one of my legs had gone to sleep.

A considerable time later my attention was once again drawn to marmosets. I went down to an animal dealer's shop in London, to inquire about something quite different, and the first thing I saw on entering the shop was a cage full of marmosets, a pathetic, scruffy group of ten, crouched in a dirty cage on a perch so small that they were continually having to jostle and squabble for a place to sit. Most of them were adults, but there was one youngster who seemed to be getting rather a rough time of it. He was thin and unkempt, so small that whenever there was a reshuffling of positions on the perch he was always the one to get knocked off. As I watched this pathetic, shivering little group, I remembered the little family party I had seen in Guiana, grubbing happily for their dinner among the orchids, and I felt that I could not leave the shop without rescuing at least one of the tiny animals. So within five minutes I had paid the price of liberation, and the smallest occupant of the cage was dragged out, screaming with alarm, and bundled into a cardboard box.

When I got him home I christened him Pavlo and introduced him to the family, who viewed him with suspicion. However, as soon as Pavlo had settled down he set about the task of winning their confidence, and in a very short time he had all of us under his minute thumb. In spite of his size (he fitted comfortably into a large teacup), he had a terrific personality, a Napoleonic air about him which was difficult to resist. His head was only the size of a large walnut, but it soon became apparent that it contained a brain of considerable power and intelligence. At first we kept him in a large cage in the drawing-room, where he would have plenty of company, but he was so obviously miserable when confined that we started letting him out for an hour or two every day. This was our undoing. Very soon Pavlo had convinced us that the cage was unnecessary, so it was consigned to the rubbish-heap, and he had the run of the house all day and every day. He became accepted as a diminutive member of the family, and he treated the house as though he owned it and we were his guests.

At first sight Pavlo resembled a curious kind of squirrel, until you noticed his very human face and his bright, shrewd, brown eyes. His fur was soft, and presented a brindled appearance because the individual hairs were banded with orange, black and grey, in that order; his tail, however, was ringed with black and white. The fur on his head and neck was chocolate brown, and hung round his shoulders and chest in a tattered fringe. His large ears were hidden by long ear-tufts of the same chocolate colour. Across his forehead, above his eyes and the aristocratic bridge of his tiny nose, was a broad white patch.

Everyone who saw him, and who had any knowledge of animals, assured me that I would not keep him long: marmosets, they said, coming from the warm tropical forests of South America, never lived more than a year in this climate. It seemed that their cheerful prophecies were right when, after six months, Pavlo developed a form of paralysis and from the waist downwards lost all power of movement. We fought hard to save his life while those who had predicted this trouble said he ought to be destroyed. But he seemed in no pain, so we persevered. Four times a day we massaged his tiny legs, his back and tail with warm cod-liver oil, and he had more cod-liver oil in his special diet, which included such delicacies as grapes and pears. He lay pathetically on a cushion, wrapped in cotton-wool for warmth, while the family took it in turn to minister to his wants. Sunshine was what he needed most, and plenty of it, but the English climate provided precious little. So the neighbours were treated to the sight of us carrying our Lilliputian invalid round the garden, carefully placing his cushion in every patch of sunlight that appeared. This went on for a month, and at the end of it Pavlo could move his feet slightly and twitch his tail; two weeks later he was hobbling round the house, almost his old self again. We were delighted, even though the house did reek of cod-liver oil for months afterwards.

Instead of making him more delicate, his illness seemed to make him tougher, and at times he appeared almost indestructible. We never pampered him, and the only concession we made was to give him a hot-water bottle in his bed during the winter. He liked this so much that he would refuse to go to bed without it, even in mid-summer. His bedroom was a drawer in a tall-boy in my mother's room, and his bed consisted of an old dressing-gown and a piece of fur-coat. Putting Pavlo to bed was quite a ritual: first the dressing-gown had to be spread in the drawer and the bottle wrapped in it so that he did not burn himself. Then the piece of fur-coat had to be made into a sort of furry cave, into which Pavlo would crawl, curl up into a ball and close his eyes blissfully. At first we used to push the drawer closed, except for a crack to allow for air, as this prevented Pavlo from getting up too early in the morning. But he very soon learned that by pushing his head into the crack he could widen it and escape.

About six in the morning he would wake up to find that his bottle had gone cold, so he would sally forth in search of alternative warmth. He would scuttle across the floor and up the leg of my mother's bed, landing on the eiderdown. Then he would make his way up the bed, uttering squeaks of welcome, and burrow under the pillow where he would stay, cosy and warm, until it was time for her to get up. When she eventually got out of bed and left him, Pavlo would be furious, and would stand on the pillow chattering and screaming with rage. When he saw, however, that she had no intention of getting back to bed to keep him warm, he would scuttle down the passage to my room and crawl in with me. Here he would remain, stretched luxuriously on my chest, until it was time for me to get up, and then he would stand on my pillow and abuse me, screwing his tiny face up into a ferocious and most human scowl. Having told me what he thought of me, he would dash off and get into bed with my brother, and when he was turned out of there would go and join my sister for a quick nap before breakfast. This migration from bed to bed was a regular morning performance.

Downstairs he had plenty of heating at his disposal. There was a tall standard-lamp in the drawing-room which belonged to him: in the winter he would crawl inside the shade and sit next to the bulb, basking in the heat. He also had a stool and a cushion by the fire, but he preferred the lamp, and so it had to be kept on all day for his benefit, and our electricity bill went up by leaps and bounds. In the first warm days of spring Pavlo would venture out into the garden, where his favourite haunt was the fence; he would sit in the sun, or potter up and down catching spiders and other delicacies for himself. Halfway along this fence was a sort of rustic arbour made out of poles thickly overgrown with creepers, and it was into this net of creepers that Pavlo would dash if danger threatened. For many years he carried on a feud with the big white cat from next door, for this beast was obviously under the impression that Pavlo was a strange type of rat which it was her duty to kill. She would spend many painful hours stalking him, but since she was as inconspicuous as a snowball against the green leaves she never managed to catch Pavlo unawares. He would wait until she was quite close, her yellow eye glaring, her tongue flicking her lips, and then he would trot off along the fence and dive in among the creepers. Sitting there in safety, he would scream and chitter like an urchin from between the flowers, while the frustrated cat prowled about trying to find a hole among the creepers big enough for her portly body to squeeze through.

Growing by the fence, between the house and Pavlo's creeper-covered hide-out, were two young fig-trees, and round the base of their trunks we had dug deep trenches which we kept full of water during the hot weather. Pavlo was pottering along the fence one day, chattering to himself and catching spiders, when he looked up and discovered that his arch-enemy the cat, huge and white, was sitting on the fence between him and his creeper-covered arbour. His only chance of escape was to go back along the fence and into the house, so Pavlo turned and bolted, squeaking for help as he ran. The fat white cat was not such an expert tight-rope walker as Pavlo, so her progress along the fence top was slow, but even so she was catching up on him. She was uncomfortably close behind him when he reached the fig-trees, and he became so nervous that he missed his footing and with a frantic scream of fright fell off the fence and straight into the water-filled trench below. He rose to the surface, spluttering and screaming, and splashing around in circles, while the cat watched him in amazement: she had obviously never seen an aquatic marmoset before. Luckily, before she had recovered from her astonishment and hooked him out of the water, I arrived on the scene and she fled. I rescued Pavlo, gibbering with rage, and he spent the rest of the afternoon in front of the fire, wrapped in a piece of blanket, muttering darkly to himself. This episode had a bad effect on his nerves, and for a whole week he refused to go out on the fence, and if he caught so much as a glimpse of the white cat he would scream until someone put him on their shoulder and comforted him.

Pavlo lived with us for eight years, and it was rather like having a leprechaun in the house: you never knew what was going to happen next. He did not adapt himself to our ways, we had to adapt ourselves to his. He insisted, for example, on having his meals with us, and his meals had to be the same as ours. He ate on the window-sill out of a saucer. For breakfast he would have porridge or cornflakes, with warm milk and sugar; at lunch he had green vegetables, potatoes and a spoonful of whatever pudding was going. At tea-time he had to be kept off the table by force, or he would dive into the jam-pot with shrill squeaks of delight; he was under the impression that the jam was put on the table for his benefit, and would get most annoyed if you differed with him on this point. We had to be ready to put him to bed at six o'clock sharp, and if we were late he would stalk furiously up and down outside his drawer, his fur standing on end with rage. We had to learn not to slam doors shut without first looking to see if Pavlo was sitting on top, because, for some reason, he liked to sit on doors and meditate. But our worst crime, according to him, was when we went out and left him for an afternoon. When we returned he would leave us in no doubt as to his feelings on the subject; we would be in disgrace; he would turn his back on us in disgust when we tried to talk to him; he would go and sit in a corner and glower at us, his little face screwed up into a scowl. After half an hour or so he would, very reluctantly, forgive us and with regal condescension accept a lump of sugar and some warm milk before retiring to bed. Pavlo's moods were most human, for he would scowl and mutter at you when he felt bad-tempered, and, very probably, try to give you a nip. When he was feeling affectionate, however, he would approach you with a loving expression on his face, poking his tongue out and in very rapidly, and smacking his lips, climb on to your shoulder and give your ear a series of passionate nibbles.

His method of getting about the house was a source of astonishment to everyone, for he hated running on the ground and would never descend to the floor if he could avoid it. In his native forest he would have made his way through the trees from branch to branch and from creeper to creeper, but there were no such refinements in a suburban house. So Pavlo used the picture-rails as his highways, and he would scuttle along them at incredible speeds, hanging on with one hand and one foot, humping himself along like a hairy caterpillar, until he was able to drop on to the window-sill. He could shin up the smooth edge of a door more quickly and easily than we could walk up a flight of stairs. Sometimes he would cadge a lift from the dog, leaping on to his back and clinging there like a miniature Old Man of the Sea. The dog, who had been taught that Pavlo's person was sacred, would give us mute and appealing looks until we removed the monkey from his back. He disliked Pavlo for two reasons: firstly, he did not see why such a rat-like object should be allowed the run of the house, and secondly, Pavlo used to go out of his way to be annoying. He would hang down from the arm of a chair when the dog passed and pull his eyebrows or whiskers and then leap back out of range. Or else he would wait until the dog was asleep and then make a swift attack on his unprotected tail. Occasionally, however, there would be a sort of armed truce, and the dog would lie in front of the fire while Pavlo, perched on his ribs would diligently comb his shaggy coat.

When Pavlo died, he staged his deathbed scene in the best Victorian traditions. He had been unwell for a couple of days, and had spent his time on the window-sill of my sister's room, lying in the sun on his bit of fur-coat. One morning he started to squeak frantically to my sister, who became alarmed and shouted out to the rest of us that she thought he was dying. The whole family at once dropped whatever they were doing and fled upstairs. We gathered round the window-sill and watched Pavlo carefully, but there seemed to be nothing very much the matter with him. He accepted a drink of milk and then lay back on his fur-coat and surveyed us all with bright eyes. We had just decided that it was a false alarm when he suddenly went limp. In a panic we forced open his clenched jaws and poured a little milk down his throat. Slowly he regained consciousness, lying limp in my cupped hands. He looked at us for a moment and then, summoning up his last remaining strength, poked his tongue out at us and smacked his lips in a last gesture of affection. Then he fell back and died quite quietly.

The house and garden seemed very empty without his minute strutting figure and fiery personality. No longer did the sight of a spider evoke cries of: 'Where's Pavlo?' No longer were we woken up at six in the morning, feeling his cold feet on our faces. He had become one of the family in a way that no other pet had ever done, and we mourned his death. Even the white cat next door seemed moody and de-pressed, for without Pavlo in it our garden seemed to have lost its savour for her.

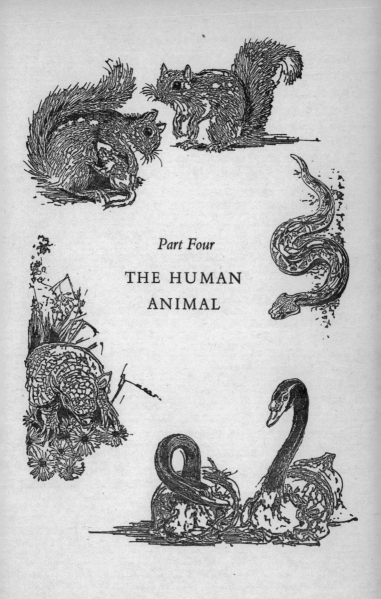

Part Four

THE HUMAN
ANIMAL

WHEN you travel round the world collecting animals you also, of necessity, collect human beings. I am much more intolerant of a human being's shortcomings than I am of an animal's, but in this respect I have been lucky, for most of the people I have come across in my travels have been charming. In most cases, of course, the fact that you are an animal-collector helps, since people always seem delighted to meet someone with such an unusual occupation, and they go out of their way to assist you.

One of the loveliest and most sophisticated women I know has helped me cram a couple of swans into a taxi-cab boot in the middle of Buenos Aires, and anyone who has ever tried to carry livestock in a Buenos Aires taxi will know what a feat that must have been. A millionaire has let me stack cages of livestock on the front porch of his elegant town house, and even when an armadillo escaped and went through the main flower-bed like a bulldozer, he remained unruffled and calm. The madame of the local brothel once acted as our house-keeper (getting all her girls to do the housework when not otherwise employed), and she once even assaulted the local chief of police on our behalf. A man in Africa – notorious for his dislike of strangers and animals – let us stay for six weeks in his house and fill it with a weird variety of frogs, snakes, squirrels and mongooses. I have had the captain of a ship come down into the hold at eleven at night, take off his coat, roll up his sleeves and set to work helping me clean out cages and chop up food for the animals. I know an artist who, having travelled thousands of miles to paint a series of pictures of various Indian tribes, got involved in my affairs and spent his whole time catching animals and none on painting. By that time, of course, he could not paint anyway, as I had commandeered all his canvas to make snake-boxes. There was the little cockney P.W.D. man who, not having met me previously, offered to drive me a hundred-odd miles, over atrocious African roads, in his brand-new Austin in order that I might follow up the rumour of a baby gorilla. All *he* got out of the trip was a hangover and a broken spring.

At times I have met such interesting and peculiar people I have been tempted to give up animals and take up anthropology. Then I have come across the unpleasant human animal. The District Officer who drawled, 'We chaps are here to help you chaps . . . ,' and then proceeded to be as obstructive and unpleasant as possible. The Overseer in Paraguay who, because he disliked me, did not tell me for two weeks that some local Indians had captured a rare and beautiful animal which I wanted, and were waiting for me to collect it. By the time I received the animal it was too weak to stand and died of pneumonia within forty-eight hours. The sailor who was mentally unbalanced and who, in a fit of sadistic humour, overturned a row of our cages one night, including one in which a pair of extremely rare squirrels had just had a baby. The baby died.

Fortunately these types of human are rare, and the pleasant ones I have met have more than compensated for them. But even so, I think I will stick to animals.

MACTOOTLE

W HEN people discover my job for the first time, they always ask me for details of the many adventures they
assume I have had in what they will persist in calling the
'jungle'.

I returned to England after my first West African trip and
described with enthusiasm the hundreds of square miles of
rain-forest I had lived and worked in for eight months. I said
that in this forest I had spent many happy days, and during all
this time I never had one experience that could with any
justification be called 'hair-raising', but when I told people
this they decided that I was either exceptionally modest or a
charlatan.

On my way out to West Africa for the second time, I met on
board ship a young Irishman called MacTootle who was going
out to a job on a banana plantation in the Cameroons. He
confessed to me that he had never before left England and he
was quite convinced that Africa was the most dangerous place
imaginable. His chief fear seemed to be that the entire snake
population of the Continent was going to be assembled on the
docks to meet him. In order to relieve his mind, I told him that
in all the months I had spent in the forest I had seen precisely
five snakes, and these had run away so fast that I had been
unable to capture them. He asked me if it was a dangerous job
to catch a snake, and I replied, quite truthfully, that the
majority of snakes were extremely easy to capture, if you kept
your head and knew your snake and its habits. All this soothed
MacTootle considerably, and when he landed he swore that,
before I returned to England, he would obtain some rare specimens for me; I thanked him and promptly forgot all about it.

Five months later I was ready to leave for England with a collection of about two hundred creatures, ranging from grasshoppers to chimpanzees. Very late on the night the ship was due to sail, a small van drew up with screeching brakes outside my camp and my young Irishman alighted, together with several friends of his. He explained with great glee that he had got me the specimens he had promised. Apparently he had discovered a large hole or pit, somewhere on the plantation he was working on, which had presumably been dug to act as a drainage sump. This pit, he said, was full of snakes, and they were all mine – providing I went and got them.

He was so delighted at the thought of all those specimens he had found for me that I had not the heart to point out that crawling about in a pit full of snakes at twelve o'clock at night was not my idea of a pleasurable occupation, enthusiastic naturalist though I was. Furthermore, he had obviously been boasting about my powers to his friends, and he had brought them all along to see my snake-catching methods. So, with considerable reluctance, I said I would go and catch reptiles; I have rarely regretted a decision more.

I collected a large canvas snake-bag, and a stick with a Y-shaped fork of brass at one end; then I squeezed into the van with my excited audience and we drove off. At half-past twelve we reached my friend's bungalow, and stopped there for a drink before walking through the plantation to the pit.

'You'll be wanting some rope, will you not?' asked Mac-Tootle.

'Rope?' I said. 'What for?'

'Why, to lower yourself into the hole, of course,' he said cheerfully. I began to feel an unpleasant sensation in the pit of my stomach. I asked for a description of the pit. It was apparently some twenty-five feet long, four feet wide and twelve feet deep. Everyone assured me that I could not get down there without a rope. While my friend went off to look for one which I hoped very much he would not find, I had another quick drink and wondered how I could have been foolish enough to get myself mixed up in this fantastic snake-hunt. Snakes in trees, on the ground or in shallow ditches were fairly easy to manage, but an unspecified number of them at the bottom of a pit so deep that you had to be lowered into it on the end of a rope did not sound at all inviting. I thought that I had an opportunity of backing out gracefully when the question of lighting arose and it was discovered that none of us had a torch. My friend, who had now returned with the rope, was quite determined that nothing was going to interfere with his plans: he solved the lighting question by tying a big paraffin pressure-lamp on to the end of a length of cord, and informed the company that he personally would lower it into the pit for me. I thanked him in what I hoped was a steady voice.

'That's all right,' he said, 'I'm determined you'll have your fun. This lamp's much better than a torch, and you'll need all the light you can get, for there's any number of the little devils down there.'

We then had to wait a while for the arrival of my friend's brother and sister-in-law: he had asked them to come along, he explained, because they would probably never get another chance to see anyone capturing snakes, and he did not want them to miss it.

Eventually eight of us wended our way through the banana plantation and seven of us were laughing and chattering excitedly at the thought of the treat in store. It suddenly occurred to me that I was wearing the most inadequate clothing for snake-hunting: thin tropical trousers and a pair of plimsoll shoes. Even the most puny reptiles would have no difficulty in penetrating to my skin with one bite. However, before I could explain this we arrived at the edge of the pit, and in the lamplight it looked to me like nothing more nor less than an extremely large grave. My friend's description of it had been accurate enough, but what he had failed to tell me was that the sides of the pit consisted of dry, crumbling earth, honeycombed with cracks and holes that offered plenty of hiding-places for any number of snakes. While I crouched down on the edge of the pit, the lamp was solemnly lowered into the depths so that I might spy out the land and try to identify the snakes. Up to that moment I had cheered myself with the thought that, after all, the snakes might turn out to be a harmless variety, but when the light reached the bottom this hope was shattered, for I saw that the pit was simply crawling with young Gaboon vipers, one of the most deadly snakes in the world.

During the daytime these snakes are very sluggish and it is quite a simple job to capture them, but at night, when they wake up and hunt for their food, they can be unpleasantly quick. These young ones in the pit were each about two feet long and a couple of inches in diameter, and they were all, as far as I could judge, very much awake. They wriggled round and round the pit with great rapidity, and kept lifting their heavy, arrow-shaped heads and contemplating the lamp, flicking their tongues out and in in a most suggestive manner.

I counted eight Gaboon vipers in the pit, but their coloration matched the leaf-mould so beautifully that I could not be sure I was not counting some of them twice. Just at that moment my friend trod heavily on the edge of the pit, and a large lump of earth fell among the reptiles, who all looked up and hissed loudly. Everyone backed away hastily, and I thought it a very suitable opportunity to explain the point about my clothing. My friend, with typical Irish generosity, offered to lend me his trousers, which were of stout twill, and

the strong pair of shoes he was wearing. Now the last of my
excuses was gone and I had not the nerve to protest further.
We went discreetly behind a bush and exchanged trousers and
shoes. My friend was built on more generous lines than I, and
the clothes were not exactly a snug fit; however, as he rightly
pointed out, the bit of trouser-leg I had to turn up at the bottom would act as additional protection for my ankles.

Drearily I approached the pit. My audience was clustered
round, twittering in delicious anticipation. I tied the rope
round my waist with what I very soon discovered to be a slip-
knot, and crawled to the edge. My descent had not got the
airy grace of a pantomime fairy: the sides of the pit were so
crumbly that every time I tried to gain a foothold I dislodged
large quantities of earth, and as this fell among the snakes it
was greeted with peevish hisses. I had to dangle in mid-air,
being gently lowered by my companions, while the slip-knot
grasped me ever tighter round the waist. Eventually I looked
down and I saw that my feet were about a yard from the
ground. I shouted to my friends to stop lowering me, as I
wanted to examine the ground I was to land on and make
sure there were no snakes lying there. After a careful inspection
I could not see any reptiles directly under me, so I shouted
'Lower away' in what I sincerely hoped was an intrepid
tone of voice. As I started on my descent again, two things
happened at once: firstly one of my borrowed shoes fell off
and, secondly, the lamp, which none of us had remembered to
pump up, died away to a faint glow of light, rather like a
plump cigar-end. At that precise moment I touched ground
with my bare foot, and I cannot remember ever having been
so frightened, before or since.

I stood motionless, sweating with great freedom, while the
lamp was hastily hauled up to the surface, pumped up, and
lowered down again. I have never been so glad to see a humble

pressure-lamp. Now the pit was once more flooded with lamp-light I began to feel a little braver. I retrieved my shoe and put it on, and this made me feel even better. I grasped my stick in a moist hand and approached the nearest snake. I pinned it to the ground with the forked end of the stick, picked it up and put it in the bag. This part of the procedure gave me no qualms, for it was simple enough and not dangerous provided you exercised a certain care. The idea is to pin the reptile across the head with the fork and then get a good firm grip on its neck before picking it up. What worried me was the fact that while my attention was occupied with one snake, all the others were wriggling round frantically, and I had to keep a cautious eye open in case one got behind me and I stepped back on it. They were beautifully marked with an intricate pattern of brown, silver, pink, and cream blotches, and when they remained still this coloration made them extremely hard to see; they just melted into the background. As soon as I pinned one to the ground, it would start to hiss like a kettle, and all the others would hiss in sympathy – a most unpleasant sound.

There was one nasty moment when I bent down to pick up one of the reptiles and heard a loud hissing apparently coming from somewhere horribly close to my ear. I straightened up and found myself staring into a pair of angry silver-coloured eyes approximately a foot away. After considerable juggling I managed to get this snake down on to the ground and pin him beneath my stick. On the whole, the reptiles were just as scared of me as I was of them, and they did their best to get out of my way. It was only when I had them cornered that they fought and struck viciously at the stick, but bounced off the brass fork with a reassuring ping. However, one of them must have been more experienced, for he ignored the brass fork and bit instead at the wood. He got a good grip and hung on like a bulldog; he would not let go even when I lifted him

clear of the ground. Eventually I had to shake the stick really hard, and the snake sailed through the air, hit the side of the pit and fell to the ground hissing furiously. When I approached him with the stick again, he refused to bite and I had no difficulty in picking him up.

I was down in the pit for about half an hour, and during that time I caught twelve Gaboon vipers; I was not sure, even then, that I had captured all of them, but I felt it would be tempting fate to stay down there any longer. My companions hauled me out, hot, dirty and streaming with sweat, clutching in one hand a bag full of loudly hissing snakes.

'There, now,' said my friend triumphantly, while I was recovering my breath, 'I told you I'd get you some specimens, did I not?'

I just nodded; by that time I was beyond speech. I sat on the ground, smoking a much-needed cigarette and trying to steady my trembling hands. Now that the danger was over I began to realize for the first time how extremely stupid I had been to go into the pit in the first place, and how exceptionally lucky I was to have come out of it alive. I made a mental note that in future, if anyone asked me if animal-collecting was a dangerous occupation, I would reply that it was only as dangerous as your own stupidity allowed it to be. When I had recovered slightly, I looked about and discovered that one of my audience was missing.

'Where's your brother got to?' I asked my friend.

'Oh him,' said MacTootle with fine scorn, 'he couldn't watch any more – he said it made him feel sick. He's waiting over there for us. You'll have to excuse him – he couldn't take it. Sure, and it required some guts to watch you down there with all them wretched reptiles.'

SEBASTIAN

Not long ago I spent some months in Argentina, and it was while there that I first met Sebastian. He was a gaucho, the South American equivalent of the North American cowboy. Like the cowboy, the gaucho is becoming rare nowadays, for most of the farms or estancias in Argentina are increasingly mechanized.

My reasons for being in Argentina were twofold: firstly, I wanted to capture live specimens of the wild animal life to bring back for zoos in this country, and, secondly, I wanted to film these same animals in their natural haunts. A friend of mine owned a large estancia about seventy miles from Buenos Aires in an area noted for its wild life, and when he invited me down there to spend a fortnight I accepted the invitation with alacrity. Unfortunately, when the time arrived my friend had some business to attend to, and all he could do was to take me down to the estancia and introduce me to the place before rushing back to the city.

He met me at the little country station, and as we jogged down the dusty road in the buggy he told me that he had got everything arranged for me.

'I'll put you in charge of Sebastian,' he said, 'so you should be all right.'

'Who's Sebastian?' I asked.

'Oh, just one of our gauchos,' said my friend vaguely. 'What he doesn't know about the animal life of this district isn't worth knowing. He'll be acting host in my absence, so just ask him for anything you want.'

After we had lunched on the veranda of the house, my friend suggested I should meet Sebastian, so we saddled horses and rode out across the acres of golden grass shimmering in the sun, and through the thickets of giant thistles, each plant as high as a man on horseback. In half an hour or so we came to a small wood of eucalyptus trees, and in the middle of it was a long, low, whitewashed house. A large and elderly dog, lying in the sun-drenched dust, lifted his head and gave a half-hearted bark before going back to sleep again. We dismounted and tied up the horses.

'Sebastian built this house himself,' said my friend. 'He's probably round the back having a siesta.'

We went round the house, and there, slung between two slender eucalyptus trees, was an enormous hammock, and in it lay Sebastian.

My first impression was of a dwarf. I discovered later that he measured about five feet two inches, but lying there in that vast expanse of hammock he looked very tiny indeed. His immensely long and powerful arms dangled over the sides and they were burnt to a rich mahogany brown, with a faint mist of white hair on them. I couldn't see his face, for it was covered by a black hat that rose and fell rhythmically, while from underneath it came the most prolonged and fearsome snores I have ever heard. My friend seized one of Sebastian's dangling hands and tugged at it vigorously, at the same time bending down and shouting in the sleeping man's ear as loudly as he could: 'Sebastian – Sebastian! Wake up, you have visitors.' This noisy greeting had no effect whatever; Sebastian continued to snore under his hat. My friend looked at me and shrugged.

'He's always like this when he's asleep,' he explained. 'Here, catch hold of his other arm and let's get him out of the hammock.'

I took the other arm and we hauled him into a sitting position. The black hat rolled off and disclosed a round, brown chubby face, neatly divided into three by a great curved moustache, stained golden with nicotine, and a pair of snow-white eyebrows that curved up on to his forehead like the horns of a goat. My friend caught hold of his shoulders and shook him, repeating his name loudly, and suddenly a pair of wicked black eyes opened under the white brows and Sebastian glared at us sleepily. As soon as he recognized my friend he uttered a roar of anguish and struggled to his feet: 'Señor!' he bellowed. 'How nice to see you. . . . Ah, pardon me, señor, that I'm sleeping like a pig in its sty when you arrive . . . excuse me, please. I wasn't expecting you so early, otherwise I would have been prepared to welcome you properly.'

He wrung my hand as my friend introduced me, and then, turning towards the house, he uttered a full-throated roar: 'Maria – Maria – !' In response to this nerve-shattering cry an attractive young woman of about thirty appeared, whom Sebastian introduced, with obvious pride, as his wife. Then he clasped my shoulder in one of his powerful hands and gazed earnestly into my face.

'Would you prefer coffee or maté, señor?' he asked innocently. Luckily my friend had warned me that Sebastian based his first impressions of people on whether they asked for coffee or maté, the Argentine green herb-tea, for in his opinion coffee was a disgusting drink, a liquid fit only to be consumed by city people and other depraved members of the human race. So I said I would have some maté. Sebastian turned and glared at his wife.

'Well?' he demanded. 'Didn't you hear the señor say he would take maté? Are the guests to stand here dying of thirst while you gape at them like an owl in the sun?'

'The water is boiling,' she replied placidly, 'and they needn't stand, if you ask them to sit down.'

'Don't answer me back, woman,' roared Sebastian, his moustache bristling.

'You must excuse him, señor,' said Maria, smiling at her husband affectionately, 'he always gets excited when we have visitors.'

Sebastian's face turned a deep brick-red.

'Excited?' he shouted indignantly. '*Excited?* Who's excited? I'm as calm as a dead horse . . . please be seated, señors . . . excited indeed . . . you must excuse my wife, señor, she has a talent for exaggeration that would have earned her a wonderful political career if she had been born a man.'

We sat down under the trees, and Sebastian lighted a small and pungent cigar while he continued to grumble good-naturedly about his wife's shortcomings.

'I should never have married again,' he confided. 'The trouble is that my wives never outlive me. Four times I've been married now and as I laid each woman to rest I said to myself: Sebastian, never again. Then, suddenly . . . puff! . . . I'm married again. My spirit is willing to remain single but my flesh is weak, and the trouble is that I have more flesh than spirit.' He glanced down at his magnificent paunch with a rueful air, and then looked up and gave us a wide and disarming grin that displayed a great expanse of gum in which were planted two withered teeth. 'I suppose I shall always be weak, señor . . . but then a man without a wife is like a cow without an udder.'

Maria brought the maté, and the little pot was handed round the circle, while we each in turn took sips from the slender silver maté pipe, and my friend explained to Sebastian exactly why I had come to the estancia. The gaucho was very enthusiastic, and when we told him that he might be required to take part in some of the film shots he stroked his moustache and shot a sly glance at his wife.

'D'you hear that, eh?' he inquired. 'I shall be appearing in the cinema. Better watch that tongue of yours, my girl, for when the women in England see me on the screen they'll be flocking out here to try to get me.'

'I see no reason why they should,' returned his wife. 'I expect they have good-for-nothings there, same as everywhere else.' Sebastian contented himself with giving her a withering look, and then he turned to me.

'Don't worry, señor,' he said, 'I will do everything to help you in your work. I will do everything you want.' He was as good as his word: that evening my friend left for Buenos Aires, and for the next two weeks Sebastian rarely left my side. His energy was prodigious, and his personality so fiery that he soon had complete control of my affairs. I simply told him just what I wanted and he did it for me, and the more extraordinary and difficult my requests the more he seemed to delight in accomplishing them for me. He could get more work out of the peons, or hired men, on the estancia than anyone I met, and, strangely enough, he did not get it by pleading or cajoling them but by insulting and ridiculing them, using a wealth of glittering similes that, instead of angering the men, convulsed them with laughter and made them work all the harder.

'Look at you,' he would roar scathingly, 'just look at you all ... moving with all the speed of snails in bird-lime ... it's a wonder to me that your horses don't take fright when you gallop, because even I can hear your eyeballs rattling in your empty skulls ... you've not enough brain among the lot of you to make a rich soup for a bedbug. ...' And the peons would gurgle with mirth and redouble their efforts. Apart from considering him a humorist, of course, the men knew very well that he would not ask them to do anything he could not do himself. But then there was hardly a thing that he did not know how to do, and among the peons an impossible task was always described as 'something even Sebastian couldn't do'. Mounted on his great black horse, his scarlet-and-blue poncho draped round his shoulders in vivid folds, Sebastian cut an impressive figure. On this horse he would gallop about the estancia, his lassoo whistling as he showed me the various methods of roping a steer. There are about six different ways of doing this, and Sebastian could perform them all with equal facility. The faster his horse travelled, the rougher the ground, the greater accuracy he seemed to obtain with his throws, until you had the impression that the steer had some sort of magnetic attraction for the rope and that it was impossible for him to miss.

If Sebastian was a master with a rope, he was a genius with his whip, a short-handled affair with a long slender thong, a deadly weapon which he was never without. I have seen him, at full gallop, pull this whip from his belt and neatly take the head off a thistle plant as he passed. Flicking cigarettes out of people's mouths was child's play to him. I was told that in the previous year a stranger to the district had cast doubts on Sebastian's abilities with a whip, and Sebastian had replied by stripping the man's shirt from his back, without once touching the skin beneath. Sebastian preferred his whip as a weapon – and he could use it like an elongated arm – yet he was very skilful with both knife and hatchet. With the latter weapon he could split a matchbox in two at about ten paces. No, Sebastian was definitely not the sort of man to get the wrong side of.

A lot of the hunting I did with Sebastian took place at night, when the nocturnal creatures came out of their burrows. Armed with torches, we would leave the estancia shortly after dark, never returning much before midnight or two in the morning, and generally bringing with us two or three specimens of one sort or another. On these hunts we were assisted by Sebastian's favourite dog, a mongrel of great age whose teeth had long since been worn down level with his gums. This animal was the perfect hunting-dog, for even when he caught a specimen it was impossible for him to hurt it with his toothless gums. Once he had chased and brought to bay some specimen, he would stand guard over it, giving one short yap every minute or so to guide us to the spot.

It was during one of these night hunts that I had a display of Sebastian's great strength. The dog had put up an armadillo, and after it had been chased for several hundred yards the creature took refuge down a hole. There were three of us that night: Sebastian, myself, and a peon. In chasing the armadillo the peon and I had far outstripped Sebastian, whose figure did not encourage running. The peon and I reached the hole just in time to see the rear end of the armadillo disappearing down it, so we flung ourselves on to the grass and while I got a grip on its tail the peon grasped its hind legs. The armadillo dug his long front claws into the sides of the hole, and though we tugged and pulled he was as immovable as though he were embedded in cement. Then the beast gave a sudden jerk and the peon lost his grip. The armadillo wriggled farther inside the hole, and I could feel his tail slipping through my fingers. Just at that moment Sebastian arrived on the scene, panting for breath. He pushed me out of the way, seized the armadillo's tail, braced his feet on either side of the hole and pulled. There was a shower of earth, and the armadillo came out of the hole like a cork out of a bottle. With one sharp tug Sebastian had accomplished what two of us had failed to do.

One of the creatures I wanted to film on the estancia was the rhea, the South American ostrich, which, like its African cousin, can run like a racehorse. I wanted to film the rheas being hunted in the old style, by men on horseback armed with boleadoras. These weapons consist of three wooden balls, about the size of cricket balls, each attached to the other by a fairly long string; they are whirled round the head and thrown so that they tangle themselves round the bird's legs and bring it to the ground. Sebastian arranged the whole hunts for me, and we spent my last day filming it. As most of the peons were to appear in these scenes, they all turned up that morning in their best clothes, each obviously trying to outdo the other by the brilliance of his costume. Sebastian surveyed them sourly from the back of his horse:

'Look at them, señor,' he said, contemptuously spitting. 'All done up and as shining as partridge eggs, and as excited as dogs on a bowling green, just because they think they're going to get their silly faces on the cinema screen ... they make me sick.'

But I noticed that he carefully combed his moustache before the filming started. We were at it all day in the boiling sun, and by evening, when the last scenes had been shot, we all felt in need of a rest – all of us, that is, except Sebastian, who seemed as fresh as when he started. As we made our way home, he told me that he had organized a farewell party for me that night, and everyone on the estancia was to be present. There would be plenty of wine and singing and dancing, and his eyes gleamed as he told me about it. I had not the heart to explain that I was dead tired and would much rather go to bed, so I accepted the invitation.

The festivities took place in the great smoke-filled kitchen, with half a dozen flickering oil-lamps to light it. The band consisted of three guitars which were played with great verve. I need hardly say that the life and soul of the party was Sebastian. He drank more wine than everyone else, and yet remained sober; he played solos on the guitar; he sang a great variety of songs ranging from the vulgar to the pathetic; he consumed vast quantities of food. But, above all, he danced;

danced the wild gaucho dances with their complicated steps and kicks and leaps, danced until the beams above vibrated with his steps and his spurs struck fire from the stone flags.

My friend, who had driven down from Buenos Aires to pick me up, arrived in the middle of the party and joined us. We sat in the corner, drinking a glass of wine together and watching Sebastian dance, while the peons clapped and roared applause.

'What incredible energy he's got,' I remarked. 'He's been working harder than anyone else today, and now he's danced us all off our feet.'

'That's what a life on the pampa does for you!' replied my friend. 'But, seriously, I think he's quite amazing for his age, don't you?'

'Why?' I asked casually, 'how old is he?'

My friend looked at me in surprise:

'Don't you know?' he asked. 'In about two months' time Sebastian will be ninety-five.'

JERSEY WILDLIFE PRESERVATION TRUST

Gerald Durrell writes:

Have you enjoyed this book?

If you have, it is the animals that have made this possible; and these animals are not just characters in a book: they *really* exist. But many of them will not exist for much longer unless they have your help.

All over the world the wildlife that I write about is in grave danger. It is being exterminated by what we call the progress of civilization. A great number of creatures will become extinct in a very short time if something is not done, and done swiftly.

Some time ago I created on the Island of Jersey a zoological park which is now the headquarters of the Jersey Wildlife Preservation Trust. Our aim is to create a sanctuary in which we can establish breeding colonies of these threatened species, so that, even if they become extinct in the wild state, they will not vanish forever. To do this work money is required.

Therefore we need as many members as possible to join the Trust. It will cost you little, but you will be helping a cause that is of the utmost importance and urgency. I say 'urgency' advisedly, because, as you read this, yet another species is added to the danger list.

If the animals I write about have given you pleasure, please join the Trust. The animals will be greatly indebted for every subscription received.

Full particulars can be obtained from –

The Secretary
Jersey Wildlife Preservation Trust
Les Augres Manor
JERSEY
Channel Islands

MORE ABOUT PENGUINS

Penguinews, which appears every month, contains details of all the new books issued by Penguins as they are published. From time to time it is supplemented by *Penguins in Print*, which is a complete list of all titles available. (There are some five thousand of these.)

A specimen copy of *Penguinews* will be sent to you free on request. For a year's issues (including the complete lists) please send 50p if you live in the British Isles, or 75p if you live elsewhere. Just write to Dept EP, Penguin Books Ltd, Harmondsworth, Middlesex, enclosing a cheque or postal order, and your name will be added to the mailing list.

In the U.S.A.: For a complete list of books available from Penguin in the United States write to Dept CS, Penguin Books Inc., 7110 Ambassador Road, Baltimore, Maryland 21207

In Canada: For a complete list of books available from Penguin in Canada write to Penguin Books Canada Ltd, 41 Steelcase Road West, Markham, Ontario

THE BAFUT BEAGLES

Gerald Durrell

The Bafut Beagles was the name which Gerald Durrell gave to the Africans and mongrel dogs with which he hunted and captured many of the oddest and most elusive creatures in the Cameroons. His adventures in pursuit of such fauna as flying mice and booming squirrels were often as strange as the animals themselves. The Africans would not touch a Que-Fong-Goo until he provided them with a magic potion; and he did not find his first Hairy Frog until he escaped into the cold dawn after spending a night drinking with the Fon, the convivial King of Bafut, and teaching him and his councillors to dance the Conga.

The author succeeded in capturing a collection of very queer animals. In this book he has now caught their spirit and humour. A supercilious toad, a hypocritical chimpanzee, or a wide-eyed baby galago innocently dismembering a locust, is for him a character as vivid and interesting as Jacob, his remarkable cook, or even the Fon himself.

'There are not many travel books with a more natural sense of humour' – *Guardian*

THE DRUNKEN FOREST
Gerald Durrell

'Once again he has returned from his explorations with another fascinating account of his adventures. He has the ability of overcoming any ignorance of or indifference to his own subject, and he will soon have the most disinterested reader crowing with delight over the habits of orange armadillos, horned toads, Budgett's, and other endearing animals – once you get to know them.

'On this occasion it is into the Argentine pampas and the little-known Chaco territory of Paraguay where Mr Durrell and his wife go wandering off. . . . His sympathy with the animal world encourages the Disney in every creature to show itself' – *Time and Tide*

MY FAMILY AND OTHER ANIMALS
Gerald Durrell

This book is soaked in the sunshine of Corfu, where the author lived as a boy with his 'family and other animals'. It is a matter of personal taste whether one most enjoys the family, with its many eccentric hangers-on, sparring round gentle imperturbable Mother, or the animals Gerry studies and brings back to the strawberry-pink, the daffodil-yellow, or the snow-white villa. The procession includes toads and tortoises, bats and butterflies, scorpions and geckos, ladybirds, glow-worms, octopuses, and rose-beetles, Quasimodo the pigeon, the puppies Widdle and Puke, and of course the Magenpies.

The whole is presented with a blend of youthful wonder and mature artistry.

'This is a bewitching book' – *Sunday Times*

THE GOSHAWK

Terence H. White

Stories of close relationships between men and beasts – *Born Free* or *Ring of Bright Water* – possess peculiar fascination. To this T. H. White added an individuality of style and independence of philosophy which make *The Goshawk* a classic of its kind. As David Garnett has written: '*The Goshawk* is the story of a concentrated duel between Mr White and a great beautiful hawk during the training of the latter – the record of an intense clash of wills in which the pride and endurance of the wild raptor are worn down and broken by the almost insane willpower of the schoolmaster falconer. It is comic; it is tragic; it is all absorbing. It is strangely like some of the eighteenth-century stories of seduction.'

'Mr White impregnates every sentence with the fire of passion and mellows it with the tenderness of affection. I rank *The Goshawk* as a masterpiece' – Guy Ramsey in the *Daily Telegraph*

Also available

FAREWELL VICTORIA

NOT FOR SALE IN THE U.S.A. OR CANADA

MAN MEETS DOG

Konrad Lorenz

Take another look at your dog. What is he? A four-legged hearth-rug with fully automatic tail – a chip off the old jackal, in fact? Or a part-tamed wolf with the soul of a vagabond?

Take another look at yourself, too. What are you? His pack-leader? (We're joking, of course.) His mother . . . or just the man he has to buy the licence?

And how about the cat that lodges with you? Do you ever pick her up and start humming: 'Hold that tiger!'

If you read *King Solomon's Ring*, you will not have forgotten its enchanting sidelights on the behaviour of animals. The same humanity, the same expert knowledge of animals inspires this study of the relationship between men and their domestic pets, and Konrad Lorenz's pages are crammed with delightful stories from his own experience.

'His new book is the best I have read about dogs' – Raymond Mortimer in the *Sunday Times*

NO ROOM IN THE ARK

Alan Moorehead

'The best written book about big game since Hemingway's *Green Hills of Africa*' – Cyril Connolly in the *Sunday Times*

Writing principally of the spectacular wild life, seemingly sentenced by the gun and the car to the doom of the American bison, Alan Moorehead records here three journeys he made in Africa. His pages are alive with the life, the history, and the misty distances of the huge, hot continent.

'An almost cinematic vision of the country through which he travelled and of the people and animals that inhabit it . . . a very rewarding book' – *New Statesman*

'His journeys are so beguilingly recounted and his own modest candid, kindly personality so attractive that his book is a pleasure to read' – Elspeth Huxley in the *Listener*

Also available

THE WHITE NILE

DARWIN AND THE BEAGLE

NOT FOR SALE IN THE U.S.A.

THE YEAR OF THE GORILLA

George Schaller

How a man and a woman, unarmed, spent months among the mountain gorillas of the Congo.

For a year a young scientist and his wife camped in the jungle and recorded the daily movements, the social behaviour, the feeding and mating habits of man's nearest relative. The result is a factual study of nature in the wild, a book as irresistible as *Born Free* or *Ring of Bright Waters*. The legendary ferocity of gorillas makes it hard to picture them as lovable: but that is exactly how they emerge from George Schaller's extraordinary account, with its unique photographs.

'Surpasses any animal book I have read for a long time' – Gerald Durrell

NOT FOR SALE IN THE U.S.A. OR CANADA

GERALD DURRELL

Gerald Durrell is among the best-selling authors in English. His adventurous spirit and his spontaneous gift for narrative and anecdote stand out in his accounts of expeditions to African and South America in search of rare animals. He divines the characters of these creatures with the same clear, humorous, and unsentimental eyes with which he regards those chance human acquaintances whose conversations in remote places he often reproduces in all their devastating and garbled originality. To have maintained, for over ten years, the standard of *The Bafut Beagles* and *The Overloaded Ark* can only be described as a triumph.

His other Penguins are:

Three Singles to Adventure*

Three Singles to Adventure takes the reader to South America, where he meets the sakiwinki, and the sloth clad in bright green fur, where he can hear the horrifying sound of piranha fish on the rampage, or learn how to lasso a galloping anteater.
'Stuffed with exquisitely ridiculous situations' – *Spectator*

A Zoo in My Luggage

In *A Zoo in My Luggage* Gerald Durrell chronicles the birth of a private zoo. Journeying to the Cameroons he and his wife, helped by the renowned Fon of Bafut, managed to collect 'plenty beef'. Their difficulties began when they found themselves back at home, with Cholmondeley the chimpanzee, Bug-eye the bush-baby, and other founder-members, and nowhere to put them.

and

MENAGERIE MANOR

THE WHISPERING LAND

*NOT FOR SALE IN THE U.S.A.